"*Moraine* helps us understand how one's history, habits, unchallenged parental modeling, and the absence of self-examination work to keep us clinging to a culture learned but not examined, and making the same mistakes of judgment generation after generation, essentially lacking control over our own lives.

There is much here that would seem to be of enormous benefit to women, caught in the morass of inner struggles with tradition. As a man, socialized in a sexist tradition, I would add that we men also might find this a beneficial read."

H. Lawrence McCrorey
Ph.D. Professor Emeritus
University of Vermont
College of Medicine

LINDEN HILL PUBLISHING
11923 Somerset Avenue
Princess Anne, MD 21853
www.lindenhill.net

Copyright 2005

All rights reserved. No part of this publication may be reproduced, stored in a retrieval system or transmitted in any form or by any means, electronic, mechanical, photocopying, recording, or otherwise, without prior permission.

Printed in the United States

Cover art by Jeanne du Nord © 2005.

ISBN 0-9704754-6-2

Library of Congress Control Number: 2005930572

MY NAME IS MORAINE

Betty McBriety Powell, Ph.D.

LIMITED FIRST EDITION

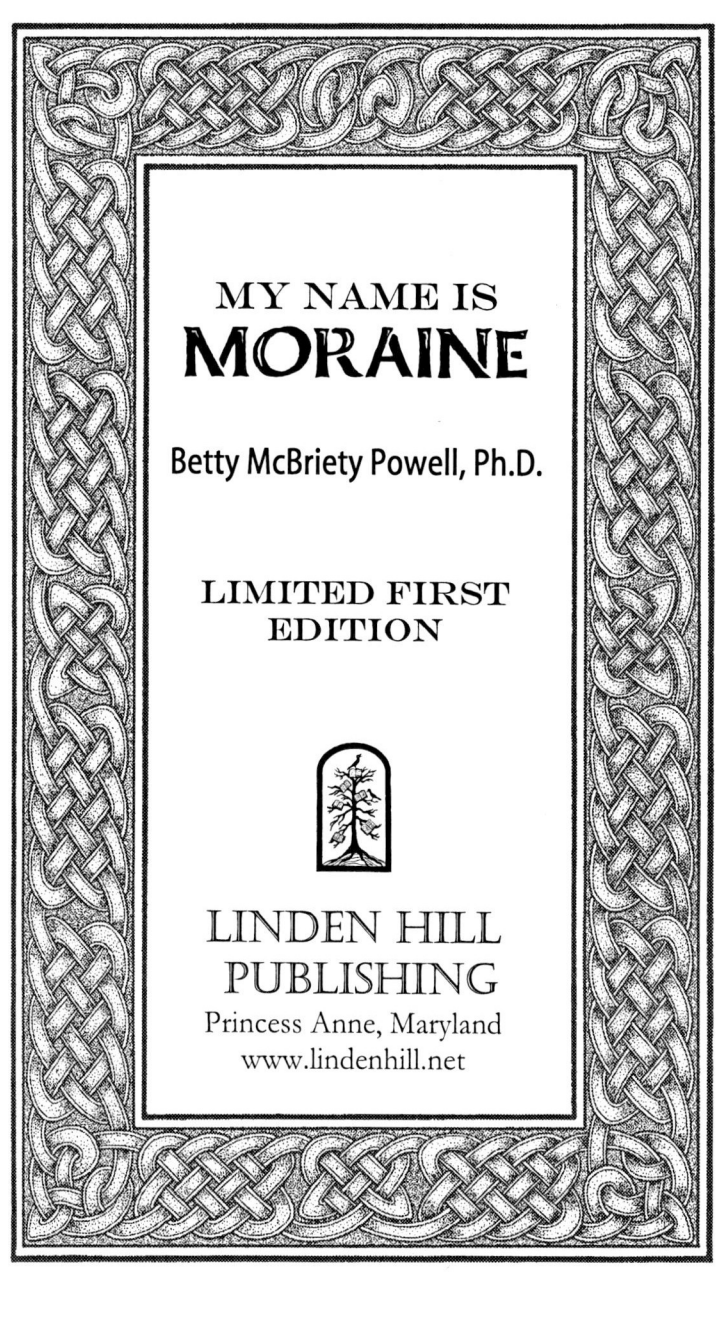

LINDEN HILL
PUBLISHING
Princess Anne, Maryland
www.lindenhill.net

TABLE OF CONTENTS

Heart of the Glacier 1

Black Shawls 5

Granny 13

Patrick O'Shea 19

Frightened Children 27

Cousin Bridget 37

Bonds of Marriage 43

From the Halls of Glory 47

Kathleen 53

Clandestine Cliques 59

Confiding Cousins 71

Little Ones 75

Off on a Jaunt 81

Picnic 85

Green, Gold, and Rosy Colours 93

Hope and Promises 97

Discord 103

Gift to the Church 113

Fathers	119
Mousy Rebel	127
Delia	133
Facing Difference	137
Meeting Ground	139
Bonaventure and Ruby	143
Night Out	151
Grannies	155
Lurking Banshee	163
St. Catherine's Wheel	167
The Glamour	175
Tears and Ice Spears	181
Wake	185
Hooters and Udders	189
Vengeance	193
Jail	197
Maskfaces	205
Epilogue	215

The old boys don't sing 'Danny Boy' at the wake of a young girl.

CHAPTER 1

Heart of the Glacier

"My name is Moraine. That's exactly what I mean, though I was christened Maureen by Father Andrew, his very own self. Moraine is a special word to me. When I was very small my father worked for a while as a fisherman in Finland, coming home only for short visits – such short visits. Although my Mother missed him most awfully and he missed the warmth of family life, there was no other work for him.

He was full of marvels to share with us children. The Northern Lights that pulsate the whole spectrum seemed to overshine even our Irish rainbows, which I had considered to be the most magical things in the universe. Then he told us about the glaciers which covered much of the land and which, even then, were beginning to respond to climatic changes. We children could not begin to fathom the depth of the blue that he tried to explain. We had thought glaciers were pure white ice but the oxygen, the life giving oxygen, that was trapped within as the freezing began was an indescribable blue. As the edges of the glaciers began to melt, the rich earth that had been beneath them began to rise from the centuries of suppression. That earth is the moraine. To me, the word speaks of an awakening, and for that reason, I have changed my name. The melt takes a lifetime.

For me to have become Moraine is like global warming, because I have finally melted away the glacier I froze over myself for so many years. For twenty years, despite perfectly normal intelligence, I lived the accepting Irish housewife role. Whether or not I was wearing the black shawl, the old darkness covered me.

On my way to the church for confession, to ask for forgiveness, I told myself that I was in need of some penance for finding it difficult to forgive Paddy for hitting me. I should not have upset him after such a hard day's work, not to mention the fact that he was drunk. My mother-in-law always told me that I didn't know how to handle him and I thought she must be right. Then, almost worse, he began ignoring me. The confessional booth gave me a place to hide (under my glacier with its frozen beauty) and kept my emotional pain under control.... for a while.

When it was no longer acceptable for me to confess, being in a mixed marriage and all, there was one priest who understood my need and secretly heard me - and absolved me. Then he left the parish for the United States and I was left with a very rigid priest who convinced me that I was going to Hell right away, not waiting to be dead.

If I had been aware of my part in the conflict right then, I could maybe have saved a ton of heartache. My mixed loyalties caused me to repeat the anger and frustrations that my female forbears had passed down. It never occurred to me that my every action was reborn in those who observed me and lived with me. My children saw in me the seeds of conflict as I reacted to my husband and to them with fury. Then that fury moved on to the constant troubles in my country. The women I know never imagined that we could have slowed and then perhaps stopped the whole mess with no more than the power of women.

Just say 'no' is more than a slogan. It is the way we must say 'no' to more senseless conflict by making better choices. What are the reasons why women cling to their belief in the conflict in Northern Ireland? Among the reasons are ignorance, tradition, loyalty, and the clinging of centuries of the same culture. These reasons are not worth the loss of one life; especially if it is your own husband or child. However, there are no lines drawn in the sand to tell us when the price of loss is too great or when we shall be named cowards.

I believed those who said that the violence in Northern Ireland, my Northern Ireland, was a necessary evil; that the fighting was for justice, that there was no other way. I believed that there were two sides to this conflict just as my sisters all over the earth believe.

I believed if I behaved in a ladylike way, no one would suspect my anger came from my lack of control over my own life, my fears and tension because of the Troubles, and my frustration at being thrown into the role of motherhood which I did not understand. I was a boiling pot with the lid fastened on, a pressure cooker."

CHAPTER 2

Black Shawls

This is Moraine's family story …
"Granny, Granny! I found another one!"
Moraine, only four years old, still Maureen, held her grimy baby hands up in triumph. Tightly grasped was a penny she had just spotted on the sidewalk by the bus stop. Every single day she found a coin. Was it because no one bothered to pick up so small a coin? Was it her lucky omen for the day? One day she saw a penny when she was crossing the street, a busy street, and she was afraid to pick it up. Another day, she saw one way under the seat in the bus. She was afraid to crawl under in case Granny saw her. Then she worried that her good luck from pennies would all drain away, cancelled out by the coins she didn't pick up.

"Ye've only found a wee bit of the English trash. Throw it down. Wait until we have Irish pennies with a lovely harp on. Be a nice girl, now." Nice. How many times would she have to hear that word? It came to mean never having a choice, doing only what had always been done.

Moraine knew better than to beg with Granny. Granny was always in some kind of a big hurry or always had some dire complaint against the world. Apparently as a young girl, she had always been told to hurry (Time is money) and now felt guilty if she did not appear to be busy. Yet, she was always smiling and joking. What did that mean? Could she have some

5

great expectation of a miracle in this world or had she just decided to grin and bear it? Presenting a cheerful facade had become ingrained in Granny.

Granny, in her once black, now blackish green skirt which hid her swollen ankles and wrinkled stockings, her wispy hair pulled back over her pink skull, with the ubiquitous black shawl pulled around her brown jumper, was negotiating the steps of the homeward-bound bus. The rattley old bus was headed back to the edge of the city. Home was still part of a row of identical buildings. Although not quite country, there was enough of a green patch to hang the wash out in the breeze. Such a marvel to see the sheets blowing white as snow, being almost sure they would not be stolen from the line. Granny and Ma were always extolling the wonders of living away, only a little, from the gritty grime of much of Belfast.

Moraine, on the other hand, was already full of the excitement of the city despite the gritty grime. All of those people and pubs and churches and stores full of must-haves; a glittering green bicycle, the perfect knapsack for school, a blue dress. Such a blue dress with tiny lace edges on the collar and sleeves. She even saw a real bed with a lacy canopy. Not like her narrow cot, mind you. Lace became the epitome of luxury to her, even better than diamonds.

She brushed the stringy fringe of hair out of her faded denim blue eyes. It would have taken her mother only a few seconds to trim the brown hair but she just never seemed to have the time - or the inclination. Glimpses of such marvels in these stores were only visible as she and her

Granny trudged to the grocers to buy only the most basic necessities, items that could not be produced in their tiny vegetable patch. Once, she always remembered, they bought four oranges. Two requirements: to be able to pay and to be able to tote, between the two of them, all of the parcels. It would have been nice to have her mother go with her instead of Granny because Ma could be coerced into buying her a candy stick and Granny wasn't giving up her pennies for such fripperies (unless it was for herself). "A penny saved is a penny earned," she intoned endlessly.

Moraine squatted quickly and slipped the dropped penny into her too big, untied shoe. She hurried to get on the bus and sat down next to Granny who was contentedly munching on a biscuit. Granny just never seemed to have enough to eat and she needed to know there was more where that came from.

Moraine's mother knew that Granny had never gotten over the fear of seeing her family all broken up and starving during the Great Hunger. Memories of the Great Hunger had been passed down just as the black shawl had been passed down. Her sisters had died trying to feed their families on nothing. Her brothers had moved to Canada. Her father just gave up and her mother gave all she had to her children until she wasted away.

It was a memory and an anger that stayed with her family for generations. No one had dared to throw away a crust of bread or an old dried up turnip for fear that there was no more around the corner. From childhood on, Granny tried to hide her frustrations under her cheery front. Only occasionally, when she found a bad onion in the bin and had to throw it away, or when she heard a sudden backfire of a lorry, did she give herself away with an expression of utter fear. Sadly, the backfiring was occasionally gunfire or car bombing.

Moraine was feeling extremely proud of herself today. She had finally inherited her older sister's best dress and shoes. All much too big, but still her own now. There had been a stain right down the front button placket and her Ma, more from fear of what 'others might think' than consideration of Moraine, had appliquéd a green striped giraffe made from a piece of another old dress. She had chosen a giraffe because his long neck reached the length of the placket hiding all of the unsightly chocolate spill. The incongruity of this creation on a pink dress printed with white fuzzy kittens never occurred to Moraine, but it is probable that the example of poor taste lasted all her life.

Moraine never got with the style of the day, never was smartly turned out, even when she went to work later in life and read all of the latest fashion magazines. She carried, however, the innate dignity and aristocracy of the Irish woman. Never a tall woman, her bearing was just haughty enough to bring lesser creatures to their knees.

Granny was a paradox. Always bread making, bed making, hoeing, toothlessly smiling, she seemed to be forever serene despite the tight cords of her wrinkly old neck.

She went on at great length about how she had promised Grandpa on his deathbed that she would carry on his cobbling business and that she would light a candle for him every day for the rest of her life. She lived up to the promise about the candle, although it became only an automatic thing after a few years, something to complain about. She would have been hard put to remember what she could have been thinking with that promise.

As far as the cobbling went, she was never skilled enough to entice customers to the tiny shop, nor was she gracious enough. The weeds grew tall and poisonous looking vines took over the one window. Eventually, the building had to be torn down because it was a menace, tempting children to climb and frantic lovers to hide there.

She was inordinately proud that her husband had not been a lowly sheep farmer even though they might not have been so everlastingly hungry if he had been. Many of the villagers where they had lived had no shoes to begin with and cared very little about in investing in such a luxury.

There was a rigidity about her responses. There seemed to be a head of steam built up somewhere within. Shockingly enough, the most striking thing to Moraine was the way Granny became after her stroke, when she needed more and more care. She began to show some of what she had been hiding underneath her facade of 'all's well.' Her dislikes included but were not limited to all Protestants (of which she had met only two) and all things English.

Ma did all she could to make the household more pleasant, smiling with that same phony smile that she had always hated in her own mother. She had only recently been through the care and death of Da's mother, which had been

9

an unwelcome episode. She couldn't get over the feeling that her husband's mother was considered just a fixture. As far as Da was concerned, the old lady, his own mother, lonesome and ignored, could have been anyone off the street. Ma was therefore the comforter and the unwilling caretaker.

She was not a hugger. She gave few hugs to her little girls, Moraine and her sister, when they really needed them. The boys got a few extra hugs just because they were boys. Does one need lessons to know when a hug is needed? She had truly loved and tried to show affection for the old woman, who was easy and kindly at first, but finally the long days and the ingratitude left Ma seething.

Granny became the complete harridan. No facade of politeness or consideration came between her and her big mouth. She had kept all of her real feelings so tamped down that, now that some of her controls were released by age and illness, she began to catch up with her emotions.

She cursed - Moraine could not have learned more from a sailor fresh from the sea - and she let the world know what she thought of her late husband's habit of drinking too much *poteen* (home brewed whiskey), his countless weaknesses, and she was not altogether sure that her daughter had not inherited the lot. She thought the Unionists were holding up her personal destiny. Not much that was British was forgivable to her.

So, where was her fresh water jug, her wee cup of tea? On and on and on she went. Ma just smiled and gave herself one luxury -- a fantasy of herself without a care, with freshly manicured nails painted an outspoken crimson. Not much of a wild and crazy fantasy but to her it was symbolic of a life of ease, when she would be answerable to no one but herself. No person who spent her days dealing with bleach, hot water or dirt could ever have the satiny, unmarked, cerise nails that she envisioned. The only time her family could remember her having a real belly laugh was at the old saw: "There will be a peter-pulling contest at St. Taffy's church on Sunday."

"Is it too much to ask, after all I've done for you, that I should have my wee cup while it's still hot? And one of the square slices of bread? Where are my teeth? Don't throw that away. I can eat it." Granny droned on.

Still Ma smiled as she slapped down the teacup, sloshing it around to be mopped up. Everything is all right. No neighbor shall say of me that I neglected my own mother. No dirty linen airing here. Her tight neck muscles and nervous hands betrayed her. She snapped at her children. She pulled her obligatory black shawl more tightly around her thinning shoulders, slapping and snapping her way along.

Then, Ma told Moraine the myths she had grown up with. She fantasized, had she lived then, she would have been a queen with power. Her wish for control in her life, a day to make choices of what to do next, seeped into Moraine's psyche.

When Moraine started to school, she felt like an outsider. She formed a façade of a proper, that is to say, a 'nice' little Catholic girl, learning her catechism, being outwardly respectful. After all, she surmised, if her feelings didn't show, they weren't there. She would have control someday.

11

As she grew up, she joined all of the other females she knew in that patriarchal world. She absorbed the tension in the air around her. She cringed at the sound of a bus backfiring. It made her want to dive for cover. She wept with the neighbor whose husband had been kneecapped for being Catholic/Protestant/Nothing. She knew there were people in ski masks who looked like the fathers and brothers of her friends in the daylight who may have been egged on by their women or at least encouraged.

Hate festered inside and she passed it on to her family. Conversations that were whispered at home were common dialogue on the schoolyard by the children who experienced the events. Embellishments made them so much more fearsome and delicious. There were tales of houses being searched, of men working in England, the only place where jobs existed for Catholics, who had another woman and how pissy drunk Oisin's father had been after the last funeral.

Their neighbor, Declan, had been killed in a car bombing. The bomb had been set in the wrong car by Declan's niece. There was more wailing about the horrible mistake than about the loss of Declan.

The wake had been a gathering place for sharing delicious tidbits and expanded anecdotes. Sometimes revenge was planned in whispered voices at these gatherings and the women approved and encouraged, even while the body lay quiet and cool with candles at each corner of the casket.

CHAPTER 3

Granny

Moraine hated seeing her Granny so helpless. She had become a bitter old woman, fighting back too late to make a change in her life. What Moraine now saw as rheumy, beady little eyes had once been bright and hopeful. Much later, Moraine would know that in old age everything sags including eyelids.

Moraine had memories of picnics and neighborhood *hooleys* where Granny had told stories and danced the step dances. How Granny had hated the dances that required her to keep her arms tightly to her sides when she wanted to be free. She was unable in her mind to glamorize dances that had been brought about by the old Penal Laws and that now were known as Irish dancing. Folks even won ribbons, cups, and prizes for dancing with their arms pinned to their sides when the legs could fly up in the air so effortlessly.

After her stroke, Granny drooled like a baby. She kept losing her teeth. When she wet her bed, she blamed it on someone spilling water. And she smelled bad. Her uniform was a lavender chenille robe with the tufts worn off and a pair of green fuzzy slippers that stank of urine from frequent accidents. It was rare that they could be removed from her to be washed. Ma could not keep up with it all. Moraine and her sister were pressed into service, but the brothers were Scot free to go and play or just lounge around, a fact that did not escape the two daughters cast into the forever role of caregivers. When they wanted to go out and play after school, the chore that had begun with love, ended up in resentment of an old lady's demands. The boys were nowhere to be seen.

Moraine hated seeing her mother helplessly stuck somewhere between the demands of Granny, the needs of her children, and the control of her husband. Later in her own life, when Moraine was a wife and mother, she too felt stuck between the demands of her mother, the needs of her children and the control of her husband. It made being a woman so undesirable, so frustrating.

When her mother started having 'nerves', as all women's health problems were termed, Moraine suspected that all women came to that end. After all, three of Ma's best friends were known to have hysterical crying fits from time to time. Her Mother did not become hysterical. She instead became more withdrawn, ignoring all but the most urgent tasks. The children were somewhat clean and that, she deemed, was enough.

Moraine's mother said over and over that she was very lucky not to have to clean at night in Belfast offices to make money like so many. Well, maybe. She could have used the money and even though the work was strictly illegal, she occasionally dreamed of being alone, scrubbing and singing away, in the quiet of the night. Moraine's father would not hear of it, of course, because the care of Granny would have fallen to him on those nights. He might lose face at the pub, though those whose wives did work at night roundly denied it. He declared pontifically that 'no wife of his would work out of home.'

When Moraine became of age to start school, she felt, for a while, as though she had been set free. She no longer had to sit with Granny, smell her, listen to her.

The nuns despaired of ever making her at least look ladylike. She was of the type who always looked as though she had rolled in the dirt before she came into the schoolroom. Her socks slipped down in her shoes and her hair was always a failure. At first, her mother had carefully rolled her hair into rags before she went to bed to set it over night into curls. Every day the rain or the mist or whatever dampness there was would conspire to straighten it in spite of that effort. She finally gave up and Moraine struggled on with her confidence-lowering stringy hair.

Moraine loved books, especially ancient history about the Celts. Because they were all girls at the parochial school, she had very little contact with the wider world of the rest of the counties of Northern Ireland and of Protestants and of boys. From the nuns she learned that there were dangers at every crossing and there was a palpable sense of tension all of the time.

One sunny day, before Granny took ill, while waiting for her to pick through every turnip in the market, Moraine was approached by three boys about her age who kept pushing her against the wall, saying, "What are you? What are you?" The redheaded one with all the freckles kept jabbing her in the shoulder, which bruised.

"You don't look like a Catholic. Are you a damn Protestant?"

Terrified, Moraine could think of no answer

15

that would apply. She thought of her identity. She was a girl, an Irish girl. She was in school. What else could she be?

At home, she asked her Ma, "What am I?" For a change, Ma took time to put her basket of wash down, pull her black shawl closer, and sit tiredly down, every bone aching. Arthritis had begun to take its toll. She looked Moraine in the eye.

"You are a good Irish Catholic colleen and I hope you stay that way. Tell them that if it ever happens again." Then, tiredly, Ma gathered her shawl, her basket of clothes and hurried off on her endless tasks. What could she tell her child that would give her an understanding that she herself lacked? Moraine had hoped for a more advisory comment, maybe a comment that would give her permission to bloody a nose or two. She would love to have poked them all in the nose, but of course, she was outnumbered and she had to be 'nice'. From this she came to believe that the whole problem was with Protestant trouble-makers though she had no idea who the boys were or 'what they were'. Her mother had been no help at all.

As a schoolgirl, she had an opportunity to make friends - all proper ones, Catholic, 'nice' naturally, but she never brought anyone home to tea. The truth was, she was ashamed of Granny and of the smell of her small home. The smell of peat and boiled cabbage might have smelled homey but the smell of urine ruined that possibility. Besides, the house was dark inside. The little old house had other houses on each side in order to keep out the cold and had been built with few windows so that even when they got electricity, the darkness prevailed. Another Grannyism: "Turn off that light! Do you think money grows on trees?"

The outdoors called to Moraine no matter how bitter the cold. She loved to have fresh air and to play all of the racing games she and her friends could make up.

Moraine also played pretend games. She pretended she lived in one of the green, green Antrim glens up the Coast Road amongst all of the rocks, and she believed she was one among all of the little people and all of the ghosts of the ancient warriors. In her fantasy, she was a queen, fully in control of all around her. It was hard to shed those images of power when she came back to the real world. When she went back to school, the other girls could not understand how she could walk so regally nor be so bossy. They accused her of thinking she was better than they. She felt that she was different than they.

Finally, she made two special friends. A triad of females is often portrayed as almost sacred, like The Three Graces. These girls were anything but sacred; they fought jealously over each other's friendship. Red Haired Meghan was sure that Black Haired Fiona liked Moraine best; Moraine was sure that Fiona liked Meghan best. Still they were friends who understood each other; they grew together, sharing confidences that no one else could ever know. Their households were very similar in size and lack of light and non-present fathers, but that is one thing they never knew about each other, because the Irish don't hold with showing their dirty linen in public or even in private. Their fear of Protestants was an underlying principle. The really important things, like favorite film stars and most hated teachers, were the focus of their lives. Most important of all was who was the cutest boy.

Meghan was forward and brassy in her attitude to the boys - all of them. Fiona kept her eyes downcast, but she was born knowing when to glance sideways with a flutter of her long lashes at one particular boy, boy-of-the-week, that is.

Moraine's nature and home life forbade her to make any forward motions and she usually felt dumpy around her friends. And she was. Born to laugh, she was so self-conscious she could hardly look up. Her giggling was confined to the company of The Graces. She needed Meghan and Fiona for her survival in her very narrow world. Believing in adventure, chivalry and bravery, and glamour most Hollywoodish, they clung to each other through the growing up years.

CHAPTER 4

Patrick O'Shea

Later, after she met Patrick, she needed Fiona and Meghan more than ever because she could never tell her parents or the nuns about the Protestant upstart. She could think of nothing else. The excitement of the "illicit" relationship keyed her to a new world, all of which was discussed with The Graces in delicious detail.

Since early girlhood, she had been informed of the sins of temptation.

"Whativer ye do, don't be making of yourself a temptation to the young lads."

"Tis a cardinal sin and will bring ye to no good end."

"The state of virginity is the highest in the sight of the Almighty."

That is, until she married. Then she was expected to be everything her man had ever dreamed of, whatever that might be.

The young lads heard nothing like this. How she would have relished being free like some of the girls with whom she went to school. How she would have enjoyed giving in to her impulses, having sex with Thomas or Will or whoever appealed to her at the moment. Or even just to know that she could. The goddesses and queens in her storytelling would have taken to their warm thighs whatever man they wished - and then discarded them at will. It just didn't work so easily for the young ladies of St. Teresa School. Her programmed conscience would never have allowed such behavior.

Moraine was so afraid of being caught in her sins, even sins of her thoughts, that she was careful never to go into any place that might have been sinful - certain pubs and places of sport, whatever that meant. She almost never saw a male beyond priests, her father, and endless uncles. Her brothers did not count because they were, in her book, non-people.

Finally, at a parish hooley, danger arrived. She was feeling especially attractive when she looked in the mirror before she left home. Her dress was dark forest green and her mother had found a lace collar in some hiding place, way back in a drawer and had even ironed it for her. "I'll not have you scorching this linen and lace. I'll do it myself." Moraine's eyes shone with anticipation, even though she knew for sure that she would sit on the sidelines all evening. Hope springs eternal.

When all of the young ladies, high collars, long sleeves and all, and the few young men who could be cajoled into coming to such an affair (the ratio was about 8 to 1), were stiffly sitting on opposite sidelines holding cups of sticky sweet punch, in the door came three of the rowdiest, swaggeringest young men Moraine had ever seen. All three had a distinct odor of ale on them and had a look of superiority on their pimply faces. It was an *I know things you will never know* look. How she yearned to be on the 'in' side of that group. The fact that her mother would just DIE added to the intrigue.

So it was that, at the age of 16, Moraine saw at once the lad of her dreams and her first Protestant in the person of Patrick O'Shea.

"Oh, if he would only notice me," she giggled to her gaggle of friends. They were properly horrified while at the same time taking a pretty close look at the interlopers themselves.

Moraine never saw anyone but him. His hair was a dark almost mahogany red, his eyes green as the sea and his charm was awesome.

Patrick never looked straight at her, but never missed a move she made.

Of course, the young men were immediately asked to leave and the ever-watchful chaperones and the priest who was overseeing the carefully chosen music escorted them out. Patrick left with a great show of jauntiness.

Keeping wayward youth off the streets came a poor second to keeping the hooley holy.

The party dragged on after that. Because of the gender mix, some of the girls acted boys' parts in the dances and the rest of the dances were in circles holding hands. Of course, this was a relief to their chaperones who had little to do if no other body parts were touching.

Patrick waited for Moraine outside of her school the next day. He had run all the way from the public school to be sure to be there when she came out. Something magnetic had happened. Much of his bravado had vanished when he saw the shy Moraine. He was without his cronies to give him courage. He became suddenly shy. When he asked her to meet him by the fountain in the park, Moraine really tried to refuse but found herself drawn to the spot. Meghan and Fiona warned her about flirting with danger. They even threatened her with pregnancy, having not the vaguest idea what might cause such a catastrophe.

Not only was she drawn to him but it was very satisfying to walk off with him while the other girls watched jealously.

Moraine and Patrick walked together, electricity zapping back and forth between them. The urgency was well nigh unbearable in that spring of their love, their graduation from school, their graduation into a different life altogether.

Patrick was to eventually become her spouse, Paddy. How could she ever have thought she loved him? Definitions change and the eye becomes clearer. Lust covers a lot of the early territory. Himself with sparkling eyes the color of the sea and with his thick wavy hair. Hers was always so stringy and straight. Often a cigarette dangled coolly from his lower lip. He seemed so strong to her with the look of a rebel in his eyes. He reminded her of her hero Michael Collins. A uniform would have completed the picture of splendor. Later, she would realize that all heroes are not as they appear and that often they had feet of clay. She found that among the great Irish heroes there was a streak of egoism, sometimes of cruelty even though they might be poetic at heart. However, in that summer, Moraine lost her heart to the mischief Patrick O'Shea projected. She loved his rolling laugh and his abundance of friends. What an exciting life they could have together! He was her knight in shining armor on a white horse, her idea of a hero. She would be carried away in a flowing white dress.

That had been excitement she would be willing to trade now for a bit of contentment. Irishmen are indeed sensitive, singing and crying, writing poetry and laughing; they just are not always very observant when it comes to the needs of their women, she realized later. Her version of hero had changed with maturity although she wasn't quite ready for the boredom of everyday life.

As long as she could remember, she had repressed fury in order to 'keep things smooth' in the family. Her one act of rebellion, all those years ago, had been marrying a Protestant, her Patrick, whom she had so adored.

The uproar caused by that heresy - the open hatred shown by her parents and his as well had been the catalyst that long prevented her from any more divisive actions. She could not tolerate conflict. That is not entirely true. She liked things to go smoothly on the surface even if that meant tolerating the seething on the inside. Many times when her hands formed fists or when she shrugged and turned away, the conflicts were there. They were just inside. She would rather kick the cat than start an argument.

It was a terrible dark day entirely when she and Patrick told her parents about their intended wedding. She could not remember now whether the clouds had been in the sky or only in her heart as she faced them. Patrick tried to shrink into invisibility, wishing he were anyplace but directly in the line of fire. As it happened he was not in the line of fire. He was ignored after the first few minutes. One incredulous stare from her parents and the searchlight's glare was trained on Moraine.

"Ma, Da, this is Patrick O'Shea, my fiance. We have been going out for six months and now we plan to marry in the spring." The awful silence that followed was almost noisy.

"Da, please talk to me. You don't know Paddy or ye'd know he's no black sinner. Ye'd know he loves me and that all this heresy thing is from the past. He has a good job. Da, talk to me. Don't shut me out. Mum, talk to him. We can't be a family like this."

"I'll not have it. Ye'll go straight to hell. I never brought you up to marry someone of his ilk. Do you realize that you'll no longer be accepted at The Church?

What do you mean that heresy is a thing of the past? It's soap to wash out your mouth that you'll be needing. I'll wash me hands of the whole caboodle. Take yourself right down to the confessional and get started on your penance right away. Ah, Blessed Mother, what shame is this!"

Then she watched her mother pull that blasted black shawl over her head, grasp her ever-present pink plastic rosary and turn her back. Moraine suspected that her mother knew the feelings that raced through her. She must have had the same feelings when she was young and not so stooped. Now she wanted Moraine to deny the very life she herself had suppressed. Couldn't she remember?

It never occurred to Moraine that her parents, well intentioned and loving, were trying to protect her from perceived dangers in the perceived sinful world, not to mention the very tangible ones of a "Mixed Marriage." She would be ostracized by her whole world, a very small world indeed. One: she might never go to Mass again nor to Confession. Two: her babies would be some kind of mixed breed, never fitting in anyplace Three: Patrick O'Shea didn't seem much like a real man to them. Why not was never apparent; maybe they just thought he was not good enough for their little girl.

If Moraine thought that the matter was settled once she survived *her* parental confrontation, she had another think coming. Patrick was amazed to find that his not-too-churchy parents were just as adamant about him marrying a Catholic girl. They, in their ignorance, envisioned their son going through some kind of mystic ritual. The grandchildren, some said, would be signed away to the Pope. His Mother had hysterics and took to her bed. His Father got roaring drunk.

With all of this fine opposition, it is no wonder that the couple was practically glued to each other. Nothing short of a lightning bolt straight from Heaven could have put them asunder.

Apparently, God was not too upset at all.

CHAPTER 5

Frightened Children

Ironically, as soon as the engagement began, Patrick used every wile possible to delay the wedding. His job was with an auto parts company that had sprung up to supply an American automobile company in Northern Ireland. The future seemed boundless for him. It seemed that 'he needed an advance in his job', 'he had a holiday planned with the boys', and 'his new job required more of his energies'.

Commitment was staring him in the face. Patrick, in his manly assurance, waited ten years before becoming officially engaged to marry Moraine. The delays were ridiculous made-up reasons and to her, the wait seemed an eternity or at least half of her lifetime. Then, in the name of all that's holy, he delayed another five years before agreeing to set the date.

Some of his old life would have to go after the wedding in order for him to be home for dinner on time, to plan a life that might include Moraine's friends giggling all over the place. That stupid Fiona never looked straight at anybody. Too many females in one place unnerved him. Good to look at but perish and forbid that they should be in his own castle. He had, however, in Moraine the best of them so a bit of a swagger crept into his mien. She could sense this and with a tinkle, almost unheard, an icicle-sized piece slid from her glacier.

During the engagement, Moraine never allowed a bit of doubt to come between them, even though there crept into her innards a small smolder of resentment. She remained loyal and hopeful throughout this travail. He could and did go out with some of the available girls while she remained in her parents' home, full of frustrations, hormones raging as much as his.

Her confidence flagged; she began to feel unattractive to everyone because he paid very little attention to her. A linen and lace collar wouldn't perform the miracle and she hadn't the money to buy new clothes. Everything was a practical drab in color so it would match her drab shoes, drab tights, and drab purse. She told herself that she could have looked quite smart had she been better decked out, forgetting that she had little taste to begin with. Forgetting, or never knowing, that fancy clothes had not drawn him to her in the first place.

All true Irishwomen were supposed to be endowed with flaming, flowing red hair. Where was hers? Had some stray Roman or Viking sneaked into her heritage? Not fair. Moraine thought about this a lot. She determined that she would never let her marriage turn into the humdrum kind like those she saw round her. She would be irresistibly glamourous forever - beginning right away. She would wear earrings everyday and stockings without 'ladders'. She would not be afraid of anything in her future, with Patrick to take care of her. Despite this insurance, she learned to use a hammer, spanner, and drill with quite a bit of proficiency since he never seemed to be there when she needed him.

When Moraine had become engaged to Patrick, she had to look for a job at least until she was wed. Her parents decreed that since she was such a "sassy young thing and wasn't she the lucky one that they did not throw her out altogether?" she must begin to pay rent and a bit of the groceries like a grown-up. She was practically thrown fresh out of school, with no training of any kind, to begin another phase of her life. The fact that she was poorly educated (make that 'not educated') for anything at all did not dawn on her at first.

She only knew that she had better not even apply for a job at the shirt factory. Not only would she soon have had her fingers sewn together from pure clumsiness with the machine, but also those jobs were reserved for Protestants. Even should she be able to sneak past that barrier - was 'Catholic' practically emblazoned on her forehead? - she would surely be ostracized.

Other jobs that seemed to be open to her simply seemed to dissolve when she applied. "Sorry, that position has been filled," became almost a litany in her mind. Finally, she heard of a *creche* attached to the parochial school she had attended, where help was needed part-time to care for the children of working mothers, and she would need only a modicum of experience. She lied her way into the position, knowing she should go immediately to confession for this, in spite of her coming rift with the church.

Brazenly, she allowed that she had baby-sat all ages from infancy on up and how much she loved children, that she never missed mass and, in general, she behaved in a very prim and proper manner, thus pulling the wool over the eyes of the nuns. To her, children were just those unnecessary beings who had not attained her degree of maturity, and she had little patience or feeling for them at all, cute though they be.

At the same time, she vowed to herself that she would take classes and learn new skills. Her interest in learning, which started then and would last the rest of her life, included reading books and attending classes on almost any subject that caught her notice - dancing, badminton, mythology, and even fencing. There were free classes at the community college and, since the price was right, she took advantage of the opportunity but chose nothing her parents would have considered practical. Philosophy didn't count; welding would have counted as practical. Cooking would have been a great benefit because she hardly knew one end of a frying pan from the other and showed little interest in the whole housekeeping thing.

Her father, who had been such an adventurous young man, had forgotten youthful exploration. He folded his hands across his generous belly, sat back, and closed his eyes. It was not his intention at this stage in his life to deal with this wayward offspring. In fact, he tried to pretend that she was invisible.

Her mother never ceased prattling on about the Church until Moraine wished, prayed, to find a room somewhere but there was never enough extra money for the rent.

She even began volunteering at the Trauma Center to keep from coming home right after work. The Trauma Center needed trained counselors but they were so overwhelmed with people in pain that they would accept a good listener to help. Whole families needed more assurance, not just the wives and mothers of those who had been killed or jailed.

Although she had stretched the truth about her baby-sitting experience, unless one counts the care for her childlike Granny, she carried the deception off famously. She had a deep sense of responsibility and an amazing natural ability to organize programs. The little ones prospered under her firm tutelage, never spoiled (nor hugged, sad to say), always clean and busy learning. Charts with hard earned gold stars covered her classroom walls between bright copies of famous paintings. She had to lay low on the paintings of naked ladies, though. They were mostly Italian masterpieces of the dear Baby Jesus with the face of an old man and His Mother with her eyes rolled to Heaven.

The room she was assigned in the school was really a brand new extension of the library and had enormous plate glass windows, floor to ceiling, that not only let in natural light but looked out onto the children's vegetable garden. When the summer days came, she planned all kinds of things for the little ones that would take them outdoors as long as the weather was nice, that is, not purely raining cats and dogs. Many a pupil when grown could remember planting carrot seed and naming the song birds in the trees thanks to Moraine.

On one such rainy day in July, she had the wee children all busily coloring in their books and making up stories and making a lot of noise with their chatter, when a car bomb went off in the street right by the garden. The bomb, apparently random, aimed to harm someone, anyone, managed quite well in its purpose. The huge windows shattered. Fortunately, the panes didn't fall inward. They disintegrated into millions of shards that fell outward into the garden, scaring the bejabbers out of everyone, including Moraine. She tried to protect, enfold, buffer, comfort, treat, de-splinter and, in general, quiet the frightened children.

Moraine shook as if the glacier that surrounded her was rumbling, like a cow in the heaves of calving. It was the birth of a new thought, *'Did I have a part in letting this happen?'*

She forced herself to be outwardly calm while she and the other teachers made sure there were no serious injuries. Then they swept and tried to get the most damaged books out of the children's reach.

A couple of the teachers were rendered nearly immobile from fear and tears. The children were afraid of the blood on their little arms and legs caused by the splintering glass. As soon as they were cleaned up and bandaged if need be, they went right on playing until their panicky parents came to get them. The parents made nervous efforts to appear calm. They were waiting for another blast, another splinter in their security.

Most of the books had been damaged. Even those that showed no visible evidence of glass were dangerous in case they did harbor some shards in their pages. They were discarded just in case.

The glowing scarlet geraniums that had flourished by the large windows were destroyed. That was Moraine's nightmare - a picture of the glorious scarlet blown into smithereens that looked like blood.

The results inflicted by that one bomb aimed at no one took its toll in more than a glass window or two. The children were traumatized, the books were lost from the collection, and Moraine was shocked and furious. The bomb was like a rock thrown into a pool with ever widening circles of affect to the parents, to the stores, to the school system.

After that awful day, the children returned to their crayons and construction paper, but three started wetting themselves and others could not be coerced into sitting at any desk near a window.

"You can't make me."

And so she could not, without using force. When she appealed to the parents, they sided with the children and threatened to take them out of her care if she persisted. There was a strange unbalanced look to the room as everyone sat or played crowded away from the new shatterproof windows.

Moraine finally solved the problem by turning the whole room around so that the pupils sat or played facing away from the windows and she stood with her back to the light, directing them in their playtime even when they were blowing bubbles. This worked fine for the children but Moraine was surprised to realize that she could feel her spine tingle, and the hair on the back of her neck rise, as she stood beside the dreaded windows.

The whole result was harrowing for her. Her hate for the 'other side' settled into a chronic belly tension. If a pencil dropped, everyone in the whole room would jump and she would shout at them, "Settle down!" Parents came in, suspiciously looking around to see if there had been any new crisis. Moraine began to feel guilty as though she had caused the bombing herself.

Moraine wondered how many of those who were causing the troubles had families and if they ever wondered what it was doing to their children. She wondered how many generations of peace it would take before that inner expectant tightness of the entrails would leave all of the Irish. She could speak peace so fluently but in her heart she wanted to get her hands, her very own hands, on the car-bomber.

Nightmares became a regular part of her life after that day. In the dreams, she was gathering children in her arms to protect them from the flying glass. In some dreams, the flying glass was rushing water, in others, there were hordes of people who would trample her if she did not get away with her charges. After the nocturnal trauma of the nightmares, Moraine was fatigued and depressed all day long. She would come to work looking haggard and worn. She was so contrary that the children avoided her when they could. They were not exactly little angels, either.

She needed someone to talk to, to communicate with. She felt no real trust in anyone she knew, even her old friends. How could one tell what phrase might be repeated to what ears, even by accident? She asked her Aunt Nell for the address of a cousin who had emigrated from Ireland to Boston in the USA, which must be the safest place in the world. She and Bridget had been playmates when they were very young and had a lot in common.

At first, after Bridget left, Moraine thought about her every day, but after time and babies happened, the thoughts grew dim until now. Moraine needed someone badly, if only as a sounding board. Maybe now she could reestablish contact with Bridget and hear about the glorious life in the USA. She could also unload some of her own depression.

CHAPTER 6

Cousin Bridget

Moraine carefully worded her first letter to her cousin. She was very cautious in the first letter not to confide too much.

Dear Bridget:
At last Aunt Nell has given me your address in Boston. To tell the truth, I didn't ask her for a while, partly because I am so envious of you for living in the USA. How long has it been now? Four, five years?

When you and I were in school, who would have imagined you would marry and move so far away? We had only dreamed of going over on our own, free and independent. Good jobs would just fall in our laps, we thought.

I'm a kind of a teacher now. I managed to con the nuns into hiring me; even lied a little about my experience, God forgive me. Because I am waiting to marry Patrick (remember my dreamboat?), I've been working with very young children in order to pay a little rent to my parents. A few of the little ones are mentally retarded. There is even a study that holds that there are an unusually high percentage of developmental problems in Northern Ireland with the belief that the ongoing tension of the mothers affects the baby. We certainly do know there is tension.

The only way I can keep at my work is the belief that my lack of formal teaching education cannot harm them at that early age. At least it's a job. I don't like it much and it pays very poorly. There'd be more tips serving in a pub but the appearance of decency is a priority, especially while I still live at home. Patrick would not approve of his fiancée serving ale. Can you believe that makes any difference to Ma and Pa? As if they never went into a tavern! They are so rooted in appearances. I'm still trying to appear 'nice' and please everybody.

I have nightmares now about shattering glass landing on children. It happened here at school from a car bomb - although the glass fell outward into the garden. In my dreams, I try to run away from the sound of the bomb but I can't. Then I am frantically trying to cover all those frightened babies with my body! It is a repeat of that awful day. It resides in me somewhere. A deep hatred began its flowering in me that day. Did you know that hate could be a flower? It is red, not like geraniums, but a deep throbbing red more like blood.

It has been going through my head to wonder what my furious face and actions afterward showed the little ones. I know their faces. Terror, unbelief, excitement, no tears. When I tried to get them to come out of the destroyed room, most would not move. They seemed to feel that if they moved from that one spot where they were unhurt, the 'awfuls' would begin again. Horses will not willingly leave even a burning stable because that's home and safety to them. The children were like that.

I'm telling you, Bridey, it really made me think about this world. I'm thinking there's more to this mess we live in. Who is teaching the ones who set the car bomb in the first place? What games did they play when they were little and what stories did they hear? Things are not what I thought. Here is a glimpse of reality, cause, and effect.

What a terrible jumble to pour out on an old friend/cousin. On our first letter, too. It must be that you are now my therapist. Distance helps, doesn't it?

The best to you all,
Maureen

No, that's not right. I'm Moraine now. I feel a change beginning in me. I am starting to look at things a little differently. Write to me when you can. M.

In a couple of weeks, Moraine was delighted to find a fat letter in the mailbox which contained photos and some words that came as a surprise to her. Her vision of the grass being greener on the other side of the fence was changed a lot. Moraine's glacier now began to thaw and a huge chunk split and fell from her shoulders. The icy shield against clarity was melting away.

Dear Moraine:

So it's Moraine now, is it? Sounds affected to me. Are you trying to be someone you aren't? Or are you trying to have a rebirth and rid yourself of all of us in your past?

Truth be told, I understand your feelings. Some of the events in my life have become clearer but so far, only by hindsight.

America is different from Ireland, in spite of the numbers of Irish here in Boston. The problem I see (and the good, as well) is that we women are just the same as were our mothers and grannies. Isn't that where I learned to be so feisty and independent? We are still expected to be all for everyone, and I am frustrated that so much depends on me.

When we first got here, David had an awful time finding a job, just because we are Irish and not expected to have any brains. I thought that was all over, but apparently it still exists behind the scenes. Maybe I should run for mayor - politics are something we Irish are trusted to do very well. Women are actually trusted to be schoolteachers, but I am not certified for that anymore than you are.

There are so many different cultures stirred together that some days I am uneasy in my own neighborhood. I miss the closeness of our one-of-a-kind communities back in Ireland (even though they were not really one-of-a-kind, more like collections of individuals).

Last week, another house nearby was cordoned off with those yellow plastic tapes, showing where an Irish woman blew her top and shot her husband. He probably needed shooting and I expect she had just had enough. I'm told that she lived with a lot of tension, some of which she brought with her from Ireland and her loyalty finally wore out. The pictures of her in the newspaper are heartbreaking. Her hair is all frizzed up, uncombed, and her eyes are dark holes surrounded by baggy skin.

I try to hide my fears from my littlest one, Marie, who is four now. Look at that photo. Isn't she a doll? I hate the damn cat named Fog that sheds gray hair on everything, but it seems to bring solace to her when her day at kindergarten goes bad or when I scream at her. Children are so vulnerable and they try so hard to be brave even when baffled about adults. Then they turn around and scream at someone else.

But tension can be contagious. When she starts to school, I'll be trudging along right beside her, just like it would have been in Belfast, only here, worrying about the black and Asian kids instead of the soldiers. I'll be ready to fight or talk back and they will be fearing us with our white faces and round eyes. Marie has one special pal. Her name is Maria – the names are how they got together. Maria is from Uruguay and because she speaks Spanish at home, she is a little slow on English, especially hearing my Irish brogue. With Marie's freckles and carrot red hair and Maria's dark skin and big eyes, they make a striking pair on the playground.

The other fear we have is the building our flat is in may be torn down. The developers want to build a whole new complex that they say will be good for the area, but actually the rents will go sky high and fancy folk will be moving into the condos. Some of our neighbors have been on the street passing out flyers asking for citizen support. To fight the zoning board? Hah! We are just kept hanging and I am so tired of it. We are trying to make a home here with some stability

Have you planned your wedding yet? What ails Patrick, always delaying things? Is he the redheaded one? He's acting a lot like an Irishman, if you ask me. You'll be marrying in the Church, I guess. Will you be wearing white? That is a sort of archaic dream that we cling to in these 'modern times'.

Give my love to your sisters and those handsome brothers of yours. You probably didn't

41

even know they are handsome. All they did was pull our hair and run like crazy when we were little.

All the best,
Bridey

P.S. Are there still the most beautiful rainbows in the world in Ireland? I think I miss them as much as I miss people or home. They seem to contain the magic I left. B.

CHAPTER 7

Bonds of Marriage

For some reason, it was not thought proper for a married lady to be in charge of school children. That doesn't make much sense now, nor did it then, when Moraine and Patrick decided to go ahead with their marriage plans. Her only job with a salary, her only bit of independence, was nipped right in the bud. She was asked to leave as soon as her impending wedding was known to the Sisters. At the same time, they got an inkling of the lies she had told them to get the job in the first place! Marrying a Protestant! Blessed Mary! And she no longer went to confession!

Moraine and Patrick stood before a justice for their marriage, destroying her dream of a church wedding with herself in a white floaty dress and all her bridesmaids gathered around her. She had long ago chosen those maidens, the two whom she truly loved, Fiona and Meghan and four whom she hoped would be green-jealous of her new estate. It just didn't happen that way. Patrick forgot to order flowers; the lady who played the organ in the parlor apparently thought it was a funeral, and neither set of parents would come for the ceremony. At age twenty-five, she was married in that same dark green Sunday-go-to-meeting dress, lacy collar, high neck, long sleeves and all, but her brown hair was loose and shiny, her eyes were like stars, and she shook like a leaf during the whole ceremony. Not just anticipation of her newly

43

married life (with all the much discussed mysteries therewith), but with the very real fear that the Great God above her would strike her with lightning very soon now for marrying a heretic.

'Oh, God, does this make me heretic, too?' Before the magistrate, as she repeated her vows, she was convinced that God had closed his great masculine eyes on her for this unforgivable transgression, and she was only able to turn to her new path at the glance of her lover.

Megan and Fiona came to the marriage dressed in their Sunday best, wanting to be sure that Moraine had friends with her in a crisis. They were sure there would be one.

They were right. They were judging from the pattern of married life that they saw around them. Drinking was the major hobby among the male factory workers, and tempers flared easily over small slights. Heaven forbid that the young wife should not be home standing by the stove when they reeled in, unless, of course, she had a good paying job of her own on a different shift. Black eyes and other bruises were a common sight. Despite Belfast weather, sunglasses were a lively business in the stores as a disguise to hide the shiners.

She was Mrs. O'Shea now. Moraine's boundaries became even more confining when she stepped into the even narrower world of being a married woman. During her engagement, she had become independent through her abilities of organization and teaching but she lost that independence immediately upon marriage. Patrick was the wage earner.

Unlike many of her contemporaries who lived in a furnished one room flat, Moraine had a house. It was a narrow row house, to be sure, where she sometimes wished she could stick her elbows out to each sidewall and stretch it out a few feet.

Any baby makes changes in the life of the mother. The arrival of one baby after another changes her life entirely.

Moraine had begun to realize during the school incident that fear is greater when one is responsible for others. Her fears, which had hovered in some shapeless form over the people in her life, were now pinpointed onto her own babies.

She could feel the maternal warmth growing inside her until it penetrated her glacial shield and spread like a blanket over her babies.

She was forever devising plans to keep them safe from the outside world. Even when the first was still an infant, she never left the pram alone for a single second, as though her presence could avoid disaster.

Her mornings started about 5 o'clock. The next ninety minutes were a building anxiety as she subconsciously awaited the rotor sounds of the regular helicopter at 7:30 on the dot, a reminder of the instability and the horrors that might occur during the day.

The helicopter came in low over her rooftop on its way to the park about a mile distant. There, English soldiers were delivered to relieve the tired guards who were taken back to their camp.

The soldiers had been sent to keep order in the area. Rather than providing a sense of security, their presence emphasized the lack of security in the area.

Moraine's tensions grew and so did her dislike of the British that had been instilled in her by her granny.

CHAPTER 8

From the Halls of Glory

Moraine was a storyteller. In the ancient tradition of the Irish, she told and retold stories of heroes and sheroes, legendary men and women who gained revenge by battle and revenge by cleverness.

"Queen Medb was angered by the insult – to think that a neighboring landowner had actually thought himself a royal person at the very level she herself proclaimed. The very idea! Another idea began to brew in her brain. For seven years, she held her grudge close to her heart. Then the method of revenge dawned on her. She would stop every wagon going into his empire. She would charge a toll to all who passed in or out of her borders. After a while, the villagers' money would be gone and she could challenge him further. Then she could start a war with her treasure chest."

Moraine sat back from the side of the bed and watched the sleeping children for a while. Her recounting of Celtic tales was a nightly ritual for her and her three children. The ritual was actually the only thing she really enjoyed doing with the children. Perhaps it was because she entered a fantasy world where she could be alone with the wee ones.

Her own favorite tale was about the seal women, the selkies, because then she could imagine herself swimming away, free, from all of Belfast and all of family responsibility. The only problem with her fantasy was that surely she would soon be in a fight with some old shark that annoyed her.

The Finn (Fion McCuhal) McCool stories were popular with the children because they could laugh and laugh at Fion's predicaments that were always brought on by himself.

In the tiny room with only a small lamp glowing, Kathleen, the first-born, looked so vulnerable. Liam, her little brother, and Brandon, the youngest, had fallen asleep, mud and all, on bed and trundle.

Kathleen's small cot was barely fitted into the space between the bureau and the trunk that was never opened because it held nothing of value. There was a pair of formal gloves, the leather moldy and splitting. There were letters in the trunk, yellowed almost beyond reading, photos of relatives and their old friends no longer known or recognized by the living family.

In the bedroom was also a battered chair that no longer could be used in the parlour. Its poor broken leg was propped on books and its previously glorious green brocade seat was now only mottled beige, ecru, tan – the blah tones much like Moraine's wardrobe.

The lamp was the real trophy. Scavenged from the living room where it had some wiring trouble, Moraine had risked electrocution to try to fix it and, sure enough, it worked in spite of nearly blowing the whole circuit. The only trouble was that the shade forevermore tipped at an angle as though it were bowing. Moraine was not a believer in bowing to anything or anybody so the lamp became the focus of her resentments. She even talked to it. "Hold your stupid head up, old lamp. Show your self-respect!" The l became a symbol to her of all the bent o women she saw with their stuffed shopping bags, scurrying through the Belfast weather.

None of this was the result of poverty, unless we speak of poverty of spirit. The chair would have been discarded long ago had Moraine only been able to come from her lethargy long enough to ask one of her brothers to throw it out or even repair it. The trunk should have been burned in its entirety without even sorting its contents. But the boys almost never came by the house because she did not ask them and her meals were not desirable. When they did drop by, in all the turmoil of greeting and chatting, she would forget to ask them, so she whined to herself and in the end, all of the undone chores added up to a sort of chaos.

Perhaps the house itself offered little inspiration for decor, conjoined to the identical houses on the right and left. Even the attic was part of everybody else's attic with no dividing wall between. On and on the long room stretched with its piles of faded draperies, broken furniture, and chipped statuettes of the Holy Family.

Once, Moraine had the best part of her Christmas decorations taken from that upstairs space. There was the tiniest possibility they had been taken by mistake. Now, she was very careful never to store her blankets and what good woolens she possessed up in that loft. Moth holes she could patch. Theft, she could not.

When she went up there, recently, to store her glass preserving jars, she saw some boxes a new owner had stowed under the eaves. It was a mighty temptation to see what wonderful contents there were. Maybe there was a great treasure. Remembering her anger at her loss, she decided to leave the neighbor that much privacy and she left the box untouched.

She was wishing she could get that old trunk up there. Maybe someone would haul it for her. Maybe, with any luck, someone would steal it.

She and Patrick had moved into this house straight after their marriage. It was rather grand for them, but his sister who had lived there previously was moving out into sweller quarters and she was willing to rent it to them for a somewhat reasonable rent. She left right much behind, things that were not good enough for her new house but not good enough either to have been left in the old one.

For all its problems, the house had provided Moraine and Patrick room they could never have afforded for the babies as they came.

The cranberry red front door with no doorknob opened directly onto the sidewalk. It could only be opened with a key or by someone inside who heard the bell. Or it could be opened by force. The back door was near the gate where the dustbins were put out for collection. There was a narrow yard, not to be called a garden because it was concrete paved, and the clotheslines were usually full of clothing and sheets and towels. To give the space a garden feeling, Moraine had hung pots of fuchsias, begonias, and her favorite scarlet geraniums, all of the brightest colors, against the battleship gray of the building.

Moraine's tremendous love for flowers and plants removed her from 'the drabs'. Her hanging garden was visible from the kitchen window and cheered her up many a time. Moraine considered it to be essential but Patrick considered this money to be ill spent. Some words were exchanged over this.

Moraine's idea of romance had been altered radically with the birth of her first baby. She found she had less and less energy for Patrick, and he came home later and later. Often in his cups, he would come home prepared to disagree with everything and anything and unprepared for his conjugal duties. The first time Patrick slapped Moraine, she slapped him back. She found this to have been a mistake. Physically more powerful, he gave her a really resounding smack that threw her against the wall where she slid down on her butt. In later days, when the base of her spine hurt, she could remember that day as though it had just happened. She could also remember how aghast Patrick had been when he realized what he had done. Their lovemaking was tempestuous in the making-up and their second child was conceived.

The next day, sorting through her bruises and feelings, Moraine felt her deep, deep anger. At the pain? No. At the loss of control she had experienced and allowed? Yes. She was never the same, never as vulnerable after that. She stored up her fury for later. The anger was stored but it was not hidden. There was a tautness of her posture, her throat muscles, and a thinning out of her soft lips.

Glaciers melt only to refreeze, then in time they melt again. With every cycle of melting, freezing, re-melting, the moraine beneath the glacier grows richer and richer.

CHAPTER 9

Kathleen

When Kathleen was only a wee lass she already had an adventurous spirit. At no time could Moraine expect Kathleen to stay put. This resulted in broken bric-a-brac, mostly to Moraine's secret delight because the whatnots had belonged to her sister-in-law.

Among the broken were endless little baby-faced ceramic children on sleds, with watering cans, with flowery garlands on their little saccharine heads. Moraine was a minimalist when it came to anything that would cause extra dusting.

Endless were the bruised knees, bloody elbows, and buckets of tears from Kathleen. Moraine had little patience with the little one, being anxious to get back to homework from whichever class she was attending at the time, or her telly or, best of all, her wonderful fantasies. Kathleen, on the other hand, wanted undivided attention and continued with her "look-Ma-no-hands" life.

Kathleen was a tiny child. Her eyes, full of mischief, were the green of her father's; her hair, to Moraine's open envy, even had a little curl or at least a bend. Moraine always dressed her in grandma-smocked dresses when they went out to the store or to mass or for a visit. Moraine wondered how her mother found the time and energy to do such needlework and she was even more surprised that she lovingly made the dresses for the child whose father, Patrick, she hated. She had found the time because she dearly loved the impish little Kathleen, mostly because she envied her free little spirit (which she openly said should be squashed). At home, Kathleen was usually dirty and dressed in jeans and T-shirts. The cutesy smocked dresses were worn only under duress and the shadow of her future rebellious personality was already beginning to show. By age three, she showed a side she imitated from Moraine. Her little face, as she played house with her dolls, would become tight-lipped, a perfect image of Moraine as she strove to keep her temper and keep the peace. When Kathleen harangued her doll family, she clenched and unclenched her tiny fists as she had seen her mother do in her attempts to avoid blowing up. Had Moraine taken much note of this activity, she would have realized she had learned those traits from her own mother. The inner conflict was passing to another generation. Not only the inner conflict but also the need for revenge was being passed. There is a fine line between holding one's rage inside and being able to dissipate it by talking it out in a form of conflict resolution that would relieve the pressure.

Patrick brought Kathleen a toy phone, a bright red one, which she treasured beyond measure and considered to be a proper cuddly toy for bedtime in spite of corners and poky places. Her voice on the phone was an exact likeness of Moraine's; banked anger, pure good humor, and some bitchiness thrown in. Even her laugh was patterned on her mother's telephone voice. It was quite fakey and known to Patrick as her 'sewing circle' laugh.

Kathleen, of course, went to parochial

school, but there was a great move on in Belfast to have the young people of differing factions meet at *hooleys*, dance together, talk with one another. Moraine would have none of it. Maybe this was because of the circumstances of her own meeting with Patrick but he couldn't see the harm. He finally convinced Moraine that everyone would profit by the new approach. Not fully convinced, but compliant for a change, Moraine acquiesced.

Kathleen went off gladly, with the feeling that she would have no limits in her future. She was free. She was not afraid of bombings and scares because, in the first place, she considered herself to be immortal, and in the second place, she had grown up in the midst of such atmosphere. Whatever fears she might have had were hidden deep down inside somewhere.

Kathleen, and later the younger children, thought that the proper response was to 'get back at', to be a part of the conflict. Resolution never entered their vocabulary.

Because of Kathleen's widening interest, Moraine found herself paying more attention to the news of the day. Many new women's organizations were working for political strength, in the hope of establishing peace. At first, she saw no point in such non-violent responses. She encouraged all who would listen to continue the revenge. When she watched the telly, she often cheered aloud when the radicals scored a hit with a bombing or a shooting.

As Kathleen grew up, her interest seemed to be grounded in the more radical political groups, because, as she said, "At least, there is some action, not just meeting after boring meeting. The administration at Stormont is going nowhere except where the politicians send it."

Kathleen was galled at the lack of female participation in the overall political scene. The few who entered the circle were exceptional, while others needed to be urged just to get together for sharing meetings. The fear of danger at home or on the streets strangled them. To entice women to get together, vans from the women's organizations had to be sent to pick them up - all to avoid danger. The danger, oddly enough, was not just from some random violent act on the street, it was often from the husband at home. The men were not pleased to have their wives out of their control. Who knew what they were up to, drinking tea and whispering, even singing.

The real value of these gatherings lay in the information given the attendees. There were talks on how to save money, how to get insurance, how to maintain health through proper diet, and sometimes, they just had fun dancing and singing and laughing.

This was too tame for 'our Kathleen' who could see no progress in these gatherings. She was not afraid of anything. She was ready to fight for her rights openly, not as she had learned from her mother and grandmother before that. She took lessons in using a rifle, learned to clean and disassemble it and the fun part, learned to aim at a target.

Moraine did not catch on right away to the direction of Kathleen's new life. The steady movement left and downward did not yet disturb or attract her attention.

Like an old dog, Kathleen began to shake

off any hovering Moraine might attempt to do. She was a sassy miss and often talked back to her mother, as she hacked away at the umbilical cord. She made no pretense of being 'nice', even after being told to be so year after year.

Her brothers, as they arrived on the scene, were just so much flotsam in her life. She helped to take care of them as babies, but could hardly wait to be free of them.

Her baby sister was not born until she was having babies of her own.

CHAPTER 10

Clandestine Cliques

Black clouds, not gray mind you, lay within touching distance of one's hand. Like multiple breasts overflowing with liquid, the huge cumuli hung waiting for the pinprick that would release all that pressure.

A rainy day in Belfast offered little welcome or comfort to strangers, no more than to those natives who ventured out. The homefolks managed to ignore the weather - at least they thought they did - in spite of feeling tired and depressed and eating too much and kicking the cat. March was the worst month because there was the wind, "a black wind that would blow you inside out," as her grandmother always said. Granny had been a great one for the overused adage and managed to evoke quite an aura of wisdom by quoting them while never uttering a thought of her own. "Handsome is as handsome does", was a great favorite of hers, and she used it often with no apparent connection to the situation at hand. "Rome wasn't built in a day" was another favorite slogan used out of context.

Watching the rain begin to smack against the windows and run down in dribbles and listening to the wind slam the shutters, Moraine felt as though she were out of rhythm with the raw elements. She gave a combined shudder and a great upheaving sigh.

She eyed the worn green velour chair by the fire, full of cats with its leg still propped on the dictionary - couldn't that Patrick do anything? She longed to sit there with a hot 'cuppa', served in her china cup with pink roses on it. Again she sighed and decided that instead of stoking the fire, when she couldn't be comfortable there anyway, she would venture into the little pub where the young, like her sixteen year old Kathleen, hung out. Moraine knew that all was not well with Kathleen and she had some wild kind of fantasy that she could convince Kathleen to come home and mend her wild ways.

Kathleen had begun to change when she was only thirteen. She had lived in the Northern Ireland atmosphere of controlled tension all of her life but was sure it couldn't touch her.

Moraine knew without being told that Kathleen was 'sleeping around' much like Moraine wished to have done when she had been that young. Now, from her perspective as a mother, it appeared to be a dangerous practice.

Moraine was dreaming of a day of family togetherness. Where does an idea like that come from? There are border collies that have the same instinct for herding peoples or animals together. It doesn't always work for mothers trying to herd, maybe never. Greeting cards extol the expectation of families doing things like dancing around a Christmas tree together or seated peacefully around a golden turkey in a very Dickensish attitude of prayerful thankfulness. They lie.

Donning all of her waterproof attire, wellies as well, she gave another great sigh and headed out into the rain and wind. The great sigh had as much to do with seeing the lace curtain move at Mrs. McBride's window and the certain knowledge on this unforgiving day that the old buzzard had nothing to do but watch her movements. Mrs. McBride should have been an international spy, thought Moraine. She needs to get a life. The worst part was that she always spoke in parentheses. "On Saturday (or was it on Friday (or the day it rained (or didn't rain))), I took a pie over to Saidie's (she's the one with the gray house (the one that the roof leaks (with the husband who won't fix it))) because she has been ill." It was more than a body could tolerate.

Just so she doesn't ask me a bunch of questions about Kathleen.

Moraine leaned down and picked up a coin that was lying on the sidewalk, remembering that her Granny would have thought it to be English trash. There were no Irish coins yet but it would surely be lucky. She took the first lap on a bus, almost empty at this hour because everyone was at work. With a deep sigh and a whispered "God between me and all harm," she sloshed in puddles for about two blocks. She avoided the barricades, the soldiers, and the smart-mouthers and arrived at the snug known as the *Kool Spot* (what a name!). She immediately spotted Kathleen in a booth with her back to the door. Moraine was literally thrown back by the noise, heat, and smoke. The jam-packed crowd was young enough not to mind.

Kathleen's girl friend, who was facing the door, looked up and said, "Here comes your Ma. She is one neat lady. She seems to understand what it is to be young and not to judge every move a body makes. So easy to talk to - not like some old woman."

"My Ma? Get off it. She pisses me off. Let me out of here." And out went Kathleen, before

61

Moraine's reddening face. Moraine held her head up high, tried to think of an excuse for being in such a place - to get out of the rain? She, outwardly calm, went over to the girl friend and sat down with her.

Although Kathleen's friend thought of Moraine as a woman of wisdom, she did not realize that Moraine's own child thought otherwise.

She had been feeling young when she left home but now, she could practically feel the wrinkles drawing up her skin, taking her youth. She ordered a Coke that she did not want and started small talk that did not fool anyone with Kathleen's friend, pretending to be interested in the latest hairdos and sports results. Finally, she slunk out the door and headed for home.

What Moraine didn't know then was that Kathleen had been high at that moment and was looking for some crack cocaine that had just started to be a need for her. Crack, of all things, the hardest to break. What could she have done had she known? She would have come up against a brick wall of defensiveness and little else. Smart mouthed, no good offspring.

It would have been most satisfying to whip Kathleen until she cringed, crawled, begged for mercy, and promised to obey, to be 'nice'. Moraine thought of this with longing although all of the shows on the telly and all of the how-to books from the supermarket insist that one should not shut a child from one's love or drive the child from the home place. That love is hard to feel. There are no rules written for parents whose hearts are torn apart by hurt and helplessness.

Because all was out of her control, she

blamed the situation on the British. Why not? With police and soldiers all over the place, why would young people want to behave and take care of themselves?

    Moraine went home from her non-encounter with Kathleen at the pub with her tail between her legs, her mind swirling with frustration. What should she have done? What could she do now? Her children were the equivalent of the cowboy movie half-breeds, belonging nowhere just as her mother had warned. Half Protestant and half Catholic, Ireland was not yet accepting of offspring from mixed marriages. Truly enough, this was a thought that rarely left her sub-conscious, with a real fear deep in her gut that trouble could come, perhaps in the form of a kneecapping for Patrick. Yet she still heard that the Troubles were not religious, just political.

    Five of Kathleen's closest cronies decided to go to a film. Moraine was so eager not to have her whining around the house all evening that she all but pushed her out the door. Kathleen happened to have a new jumper, a pale blue one, so she was more easily cajoled into going. It was the right film for them, romantic, funny, and stupid. In a lovely mood, the young people were emerging from the theatre, giggling, when a shot rang out and before their very eyes, Agnes fell to the ground. Amidst the screaming people and wailing sirens, Agnes was taken on a stretcher to hospital. The medical team did all they could to bring her to life. They succeeded, but the aftermath was far worse. She had been wounded in the spine and her legs were paralyzed. Her fate would be a wheelchair forever. A random bullet that had not even been intended for her had cast this fate.

As it happened, Agnes went on to become a very useful, confident young woman. Her studies took her to the law and in her adult life she was able to help others who had been hurt as she had been. She married and even had a baby. She never lost her fear and her anger over why it had happened to her, nor did Kathleen, her best friend. Her own mother blamed herself for the accident until the day she died. Kathleen, who was not outwardly wounded, became a victim just the same. She became increasingly bitter toward the 'other side' ... whoever that could be.

The rest of the girls and boys seemed not to have been harmed. That is, unless one takes into account Louise, who became very quiet and withdrawn, Denise, who began to have fits of temper and nightmares, or Paul who became an active member of an underground team of activists.

Moraine's daily routine was interrupted many times by a thought of injury here, a memory of nightmares there. The quick, piercing thoughts were like mosquitoes buzzing about. They would bite her and she would brush them away. There would be memory flashes of this child, that child, who had been in her care years ago at the school. Like Kathleen's friends after the shooting of Agnes, they became withdrawn, moody, had temper fits. These were the long-term damages that were not covered by such statements as 'one was killed, two were injured, the rest were unharmed.'

Kathleen eventually managed to blame Moraine for the whole mess. Moraine had only strongly suggested that Kathleen go with her friends to the film but Kathleen convinced herself that Moraine had pushed them all into going. Kathleen had to blame somebody. God wouldn't paralyze Agnes without someone being close by to accuse. Kathleen and Moraine, in their own private shells, were dealing with the thinking that held that God had a hand in all this. What good was praying?

Kathleen began subconsciously to look for support. Agnes told her that she was off the mark, but in her determined way Kathleen went first to meetings of peace committees and church committees where quiet ecumenical programs were formed to help erase the violence and to support the victims step by step.

Then she found that more to her liking were the clandestine, revolutionary, trouble making cliques. Here was intrigue and excitement. Whoever saw any results from peace meetings after 200 years? What stimulation was there in discussing elections and agreements, all accompanied by a cup of very weak tea? The whole situation struck her as weak tea.

She became ever more deeply involved and soon could not have left the organization even had she wanted to do so. She knew too many of those involved, too much of their planning. Trust was not a strong point in these cliques. She could turn in names and faces. In addition to that fact, she met Sean and fell in love.

Love? The excitement of secret meetings, the scrawny beard, the stereotypical earring (that he fondly thought made him unique) and the turtleneck black shirt all conspired to weaken the knees of Kathleen, the blossoming revolutionary.

He had no job, no visible means of support

so he cadged from Kathleen so they could have food and shelter. She knew she wanted to get away from her mother and the rest of the family. She knew she wanted to be independent and all grown up with someone of her own to love. She didn't know how to pick a man. She picked a loser.

Kathleen came home late one night after Sean had told her to get out of their miserable little room. She found her mother sitting by the dead ashes of a fire wrapped in, of all things, the despised black shawl, dozing, still hoping.

"Young lady, I want to talk to you." Unintentionally, from her loneliness and frustration, she attacked her daughter with angry words. As soon as they were out of her mouth, she knew she had made a mistake. She was angry with Kathleen for causing such worry, worried that Kathleen was messing up her life and even worried that Patrick would hear them talking and try to get into the act by giving advice. He was a great one for pontificating, with the thought in mind that the more loudly he spoke the more *smacht* (obedience) would result. She was even angry with herself for pushing Kathleen into a corner. Didn't she have a right to have anger, righteous anger?

Moraine was so determined to cure the whole problem herself that she did not admit that Patrick's heavy, angry voice coming down like Zeus might have snapped Kathleen out of her actions. She and Patrick had never quite learned to work together to solve their problems, and that lack of cooperation was Number One on their list of problems.

Kathleen was still befogged from her drugs

and her first thought was to run. She turned on her spike heels and started back out the front door. She hated confrontation as much as her mother did; yet she thought she was going to change the world.

With Moraine running after her, breathlessly and clumsily, Kathleen was soon out of sight. Where she went on that wet night remained a mystery until much later when Moraine found that she had begged her way back into Sean's hovel and arms.

It was not many moons before Kathleen decided to move in with Sean for good whether he really wanted her or not. Moraine now knew how her mother and father had felt when she married Patrick. Devastated. Pain sliced through Moraine like a cleaver. Her firstborn child, lost even before she was prepared for life. The lucky pennies weren't working so far.

Finally, quiet of a sort began to prevail. Moraine's mind stopped its pointless swirl and waited, open, as she began to weigh her options. Kathleen was gone, and if Moraine were not mistaken, she looked pregnant. Dear Jesus and Mary! What would her parents say to this? Even closer to the point, what would Patrick say? She could not bear the thought of him pontificating at her about HER daughter. "You let her run loose this way." Not to mention the ubiquitous Mrs. McBride. Moraine was full of fury just thinking of the situation even before it presented itself.

With a tiny glacier melt, Moraine began to realize that all of her fears had to do with how the problem was affecting her instead of focusing on Kathleen and her need.

After all, logically, what other people

thought could not make one iota of difference in the solution. She herself must choose her next moves wisely.

That night the glacial nightmare came again with shattering glass turned to ice that only slightly melted before she awakened. She sat up shivering in the bed with tears tracing down her cheeks. She knew she must make some positive, tangible move so she would not lose Kathleen forever.

She went out and bought a rectangular blue laundry basket that would fit a baby and she filled it with a tiny mattress, pillow, a package of newborn diapers, and a box of bottles with plastic fillers. She had to take them to Kathleen's friend's house in order to find where Kathleen lived.

Her heart nearly stopped when she saw the way Kathleen and Sean lived. The blue basket seemed too clean, too pure to put in such a place. There was nothing but a smelly old mattress and a couple of cardboard boxes. They even reused styrofoam cups. The baby would either die of every germ and bacterium that could find its way in or have the strongest immune system known to man.

Their friends looked at her suspiciously, afraid she would give their headquarters away to the enemy. After all, they were 'fighting for the good of Eire', the bunch of *goblaws* (fools).

Moraine was afraid. The chill of the previous night's dream returned. She was afraid that her grandbaby would grow up in this mistrust that smothered Northern Ireland. Kathleen accepted the gift with no expression and no apparent gratitude.

In fact, she said, "I don't want you in on this baby. This is part of my new life and I don't need you nosing around."

Moraine and Kathleen parted with no warmness, just a bare truce. Moraine could not answer; her throat was closed tight with swallowed sobs and an anger that grew from her years of frustration. She must do something, but none of this peace stuff. Her daughter had been lured away by violence and she, Moraine, must have revenge.

CHAPTER 11

Confiding Cousins

Dear Bridey:

You did not tell me how the children are doing. Do they still know they are Irish? Does Americanism rub off on them?

Do you remember Mrs. McBride who lives at the end of the street? No, of course you don't. You didn't see where we lived after our marriage. Well, she is the nosiest old bird that lives. No one comes or goes in our cul de sac without her being aware and she usually knows the reason - or finds out.

It would be easier to ignore her if I were not so sensitive about Kathleen's activities, in particular the drug scene. Up until recently, when I blew up at her, we had a sort of truce. Now, if she comes home at all it is at the most outlandish hours when she and Sean have had a bout. As I watched for her by the streetlight, I could see the curtain movement or Mrs. McB.'s face at the window as Kathleen lurched in. Doesn't the old bat ever sleep?

The next step is that Mrs. McB. tells everyone she can corner. No one really likes her, but everyone can provide a cup of tea and a biscuit in order to hear the latest, often embellished, gossip.

I've done it meself a time or two when there seemed to be a scandal too rich to keep under covers, but I only go to Polly's for my tidbits. She gets a straight line from Mrs. McBride, so it is a better place to go for information.

Dear Polly has a terrible shaking in her limbs these days. It is the onset of Parkinson's I am told, and herself with the Down's syndrome child. I wonder if there is some link between her stress and her health. There seems to be a lot of that in Northern Ireland. There's a tension that has no outlet at all. Many men drink, maybe to relieve the tension. Maybe women do too, if they can.

Having written all of that, I can say that Kathleen will stay away now and live with that no-gooder. In my thoughts about this, I am afraid I have been more concerned about what the reactions will be from Mammy, Da, and Patrick than about K's future. And about 'what other people will say.'

I'm even a little relieved for myself that she is not here. I hated our house being a battleground. In spite of that small relief, in the back of my head I'm waiting for her to come dragging home, listening for her steps or the creak of the front door.

She is getting larger all the time with the baby.

Is it true that Americans have to have screens in all their windows in the summer to keep the flying bugs out?

Love, Moraine

Dear Moraine,
    The drug thing seems to be a universal disease like violence. I am so in fear that my kids will succumb to the blandishments of some of

their peers at school. It is not enough that I am uneasy about them walking to school, it is another thing to wonder about them after they get to school. Someday I shall get my hands on one of those snotty-nosed pushers and show them what a push really is. I have tried to start a Parents' group to help keep an eye on activities but there is too much opposition to groups. Each parent is only interested in his own child. For one thing, I am sure that the children's attitudes become skewed. The chief problem seems to be teaching them right from wrong and making it stick. I wonder why it is the mother who gets the blame for any lack in that department. I would do anything to protect them but it seems to be more and more out of my control. Their main effort, young as they are, seems to be to avoid getting caught in mischief instead of making better choices to begin with.

Now, some better stuff. Our eldest daughter is a dancer and can you believe she is interested in Irish dancing? She has been in a couple of competitions lately and she has done quite well in each, even with her arms tight by her side. Was it your grandmother who hated that style? That is the only way our dancers can perform and it's a wonder they don't fall over on their sides. How can they kick so high? It makes my body ache to think of it. If we reach MY goal, she will be in Dublin for the big contest and then I can come to Belfast myself for a reunion.

I too still worry about what others will think - about everything - even though my mother is not here to check on me! I also still worry about everything. A pure worry wart, as she used to say. She was a fine example. She worried about everything from hunger to being thrown into the street and having no home. 'Tis a wonder she did not do more drinking than she did and I thought she did aplenty. She even worried when she had a hangnail.

We are going to a parade tomorrow. Any excuse for a parade is a good excuse. They seem to be held here without threatening anyone, no aggressive drumming on the *lambegs*! I'm watchful, though. And we shall take cousin Iris who has just arrived from Derry and has moved in with us. She will be half-frightened to go with us and Lord help us if a car backfires. She'll freak. She should only be here a short while for which I am mightily pleased, but I have to admit that the sound of the old brogue is a good thing and makes me a little homesick.

       Love,
       Bridget

CHAPTER 12

Little Ones

Off Kathleen and Sean went every night to cellar meetings of their secret society until the two babies came within ten months of each other. Then they dragged them along to the smoke filled cellar meetings, dangerous in every way. Two of them, born so close together, and without even the blessing of the church - such a disgrace.

Moraine had a hard time holding her head up around the women who used to be her fellow guild members until she found out that children born out of wedlock were not really a unique situation. It was merely an archaic term.

In fact, it seemed that the whole generation of youngsters had decided against marriage but they needed to learn a lot about birth control. It was right there and acceptable except to most diehard Catholics.

Contraceptives were available in Northern Ireland because of the Protestant majority. The tradition for years had been that young women remained independent and young men waited until much later to marry.

Now look at Kathleen and Sean - no thoughts of consequences, those two.

Never clean, never properly fed, the children traveled with Kathleen and Sean from pillar to post. The children were so neglected that Moraine and Patrick called a halt and took them into their home. There was a rousing good debate about that. Neither Moraine nor Pat really wanted their routine further upset by this pair of feral kittens, Maeve and Kevin. Moraine and Pat were each convinced they could do a better job rearing those than any outsiders. They convinced each other that it must be done by them.

Maeve and Kevin were not in fact orphans, yet orphaned they were because Kathleen and Sean had become so caught up in themselves. They said that they were working for the freedom of Northern Ireland from the aegis of England.

Moraine's whole family, grandparents included, had the classic 'mixed marriage identity crisis'. Maybe Kathleen, because of a strong sense of wanting to belong, had chosen the Sinn Fein side. Moraine privately thought it was only a matter of having a social circle with a large dash of intrigue thrown in. Come to think of it, it wouldn't be so bad to be in the midst of some excitement herself. She underestimated their devotion to the Cause. They, in the way of Crusaders, really thought they were in the right place, doing the right thing.

It was a grey day altogether when Moraine and Patrick, Kathleen and Sean confronted each other to make the decision to take over the care of the little ones. They tried to talk about it when the children were in school, hoping to spare the children some of the pain that would surely come as they changed schools, missed their parents, learned the more strict and, no doubt, archaic ways of the grandparents.

The burden on Moraine would be tripled. She herself was now expecting her own wee one. She felt like an old lady already.

Kathleen had been instantly on the defensive in her sassy way. "You are telling me that we are not fit parents, then?" Sean, her would-be live-in (he came and went in her life a good deal), just sat there, large circles under his eyes from the late nights. Fatherhood didn't suit him. He totally missed the idea that he had sired these pups and certainly had the responsibility for them. He took no financial responsibility, seeing none. He spent no thought or affection on them. Yet the children clamored for 'Daddy'.

Moraine slipped away to a dream world inhabited by only beautiful sights. She tried to picture pretty little babies all clean and sweet smelling but they all ended looking like those damn little ceramic figures with which her sister-in-law had cluttered the place. It was too much.

Kathleen's voice came to her from a great distance, raised in intensity, all of her justifications pouring out in a torrent. Kathleen looked to her Sean and her father for support. It seemed to Moraine that she protested too much. It also seemed to her that, far back in the corner of Kathleen's eye, there was a tiny spark of relief offered by the independence once concern for the children was removed.

It appeared both Kathleen and Sean were caught up in a trend from which they had no wish to be removed. The excitement of the conflict and making Ulster fit for humans held them like a magnet.

Moraine had tried to teach Kathleen and her brothers Brandon and Liam the way to behave by feeding them well on Old Testament Christian Bible stories of the consequences of violence and injustice. In addition, she soaked their souls with all the old ways when goddesses and gods ruled and conflict was the only way.

They soaked up very little of the genuine passion. They just sang loudly and convincingly the songs about the Easter Rising and the ones about Bobby Sands, their hero. They sang without the depth of understanding that should have come with the knowledge. They would have been unwilling to deliberately starve had that come with it. They saw Michael Collins as a logical person and a real patriot, not admitting that one man's patriot is another's terrorist. His glamour overrode all.

Moraine and Patrick tried to speak reasonably though they did not feel reasonable.

"Do you know why you are going to these meetings? Have you any idea what you are trying to accomplish?"

"Of course we do. We are the real Irish patriots. We are upholding the traditions of the past ages that we've heard about all our lives. Real justice only comes from getting even. We are seeing real heroes now who aren't afraid to stand up for our beliefs even if we have to set a bomb."

Parroted talk heard at their meetings, recited by heart.

Moraine it was who had filled Kathleen's head with such and it was she who had tried to teach her what a woman really is, before she even knew her own self.

While she was praying and saying her beads to the Virgin, she was relating to the lusty, cruel goddess in the old stories. When she was telling Kathleen and her brothers to turn the other cheek, she really wanted them to strike out and win whatever battle they were fighting that day.

In those days, when the children were small, she was not only under her glacier of unawareness but she was sending mixed messages she didn't even understand herself.

Women can only rear their children at a level of understanding that they themselves have attained, so the young are often raised in the immaturity of their mothers.

Moraine warmed with the realization of this thought. A new floe melted, cracked, and left the glacier. She stood up and settled the matter once and for all by saying, "That's enough. There is a limit to what needs to be said. They are going home with me."

Patrick hung back, seeing years ahead and his own new child yet to be born. He saw endless expenses and responsibility.

He had to admit to himself that Moraine was right though. Then, to the consternation of Kathleen and Sean, he laughed because she was so damn militant about it.

Out the door she flounced and Patrick followed her.

The next day Moraine and Patrick not only had their own child on the way, but two more children to rear.

CHAPTER 13

Off on a Jaunt

Spring always seemed late and miraculous in Belfast. When it finally came, Patrick borrowed his cousin's old rattletrap of a car and announced that they were going on what he called a 'trek'. He proposed to take the two grandchildren and Moraine for a picnic and he was pretty grandiose in his talk with his cronies.

"Me and the family are taking a little jaunt up the Coast. Just a little break and a chance to all be together."

Moraine looked around for a picnic basket. They had picnicked so little in their lives that what she finally found was her grandmother's old one that had developed so many weak places that she had to line it with an ancient raincoat. When she started to clean it, she found among the mildewed rotten napkins, a tiny tarnished silver pickle fork. Tears came to her eyes as she remembered her Granny using real blue dishes and making a real festive occasion of an outing. Each person had his own special mug for soup and then after, for the obligatory tea served with the most delectable biscuits. Moraine always hated the dishwashing that was her job when the party was over.

Another memory came. She recalled her Granny slamming the food into the basket with real anger because her grandfather had come home with too much *poteen* under his belt to be able to drive them.

Granny was not very nice to the children that day. She was full of repressed fury as she started one of her house-cleaning frenzies, making Moraine's mother and all of the children hop to the effort and Heaven forfend they should slack in their endeavor.

On August 12, the Catholic Marching Day, all of Granny's energies went into planning for the parade and urging everyone to march with vigor and to be as troublesome as possible.

But for Moraine and the family today, the outing would get underway because Patrick had borrowed the car, it was already out front, and he was ready to use it. They had agreed to head up the Antrim Coast Road, only forty miles north of Belfast, so the children could see the nine glens down the middle of the county of Antrim. Moraine began to envision the incredible green of the fields, split by rocks of all sizes and shapes and the old dolmens and standing stones from the stories. She was sure in that childlike part of her heart that the 'wee folk' would be spotted in one of the glens. After all that is where they all live.

She anticipated going up the coast, through the glens, toward Ballycastle. She reveled in the prospect of the sight and sound of the sea dashing against rocks. Sometimes the rocks had ruins of an old castle or a friary still clinging precariously. Sometimes they just looked like an ancient ruin. One could hardly climb to such a place and yet monks and clan chieftains had lived in those cold stones, often for protection from invasions, or from each other. Thinking of the enormous isolation of it gave her thrills of fear and delight.

It had been a long, long time since she had been to the sea, but for some reason, she felt a true affinity for those crashing waves, the little tidal pools in the rocks with anemones and other small sea animals in them. She would really have liked to climb into an old *curragh* and head out to one of the rocky islands dotting the coast, maybe Rathlin far to the north. She had heard that there were puffins there, with their faces that seemed to say, 'I just put this beak on because I saw you coming'. Maybe she was really a seal woman returning to the sea after a torrid romance with a human.

Moraine had heard of one island where the last inhabitant had been carried off in her coffin, leaving her cottage intact, lace curtains, musty old clothes, dishes in the sink. She dreamed of going there and deciding never to come back. She saw herself sweeping away cobwebs, opening the small window to let the sea breezes in. She wanted to scrub and paint and put an old wooden bench out by the front door next to a pot of geraniums. Then she would place a shell, a whelk, on the windowsill and it would be hers. She would find beach glass and line it up so the sun would make it look like jewels, amber and green and blue.

Would Patrick ever agree to move? Would he move away from his cronies? Here there would be no more Troubles and the accompanying tensions. She would not have to worry about their mixed marriage and the threat of kneecapping. What fantasy!

The white of the pounding surf reminded her of her mother's snow-white sheets hanging on the line. Moraine had never been able to achieve that degree of whiteness. When, as a little girl, she ran through her mother's clothesline, through the sweet, clean smelling, dewy sheets, her Ma had scolded her and smacked her with the back of her hand if she could be caught. It had been worth it for that feel of being washed throughout. Now, as she dreamed of the whiteness of the gulls and the whiteness of the breaking surf, she felt washed again. It was like a baptism or a private retreat.

    She had so looked forward to this picnic, off to the Glens. While packing, she had reminded herself not to forget the saltshaker. It was the one day when Liam and Brandon had to be at school while Patrick had a holiday. It would be perfect to be off on a jaunt with just the grandchildren, Maeve and Kevin.

CHAPTER 14

Picnic

While she was assembling the thermos bottles that would serve for the tea and the soup, the little sponge cake, and thick slices of bread with honey and boiled eggs, Moraine was dreaming away. A knock came at the door.

Old Lady McBride, who usually walked right in when her curiosity brought her to the O'Shea's, was standing outside proffering a pot of jam as a ticket for entry. Of course, she had spotted the car out front and was absolutely destroyed with the need to know all about this untoward event.

Mrs. McBride hoped, in her heart of hearts, that something tragic was unfolding in the O'Shea household. Too little excitement had enlivened her in recent weeks- months, even. That tart, Kathleen, had not been in the picture lately. What was going on there?

Mrs. McBride had garbed herself for this occasion in her Sunday best. There are shades of brown that never should be in clothes and her dress happened to fall into that category. In case there had been a death or an illness in the O'Shea family, she had also donned her mink neckpiece that looked like Mrs. Franklin Roosevelt's except for the bald spots. Her splendid if sadly outdated hat spoke of the seriousness of the occasion. The remnants of a complete bird leaned forward of the brim like a ship's figurehead. This gave her the appearance of one about to soar.

Moraine wished she would.

"And how in the world are you and all your wee ones?" Mrs. McBride's bird-like features mirrored those of the bird on her hat. "I hope all is well" she lied.

Her habit of peering around while she spoke had hardly endeared her to any of her neighbors, especially to Moraine. She missed nothing and even made Moraine look at the house with more critical eyes. '*Whisht! Dust on the windowsills!*' she thought.

"Are we going on an unexpected journey?" she cooed, eyeing the old, ratty picnic basket open on the floor. Next to it, to make matters worse, was an even rattier looking suitcase. The suitcase gave the look of an escape plan although it was to be used for a blanket and some towels.

Moraine was thinking rapidly. She wanted her answer to be as cryptic as possible because, although it didn't really matter that they were planning some pleasure, she didn't want to let Mrs. McBride in on it. It was none of her flaming business.

"No. Nothing unexpected at all, nothing at all," she replied, feeling very pleased with her noncommittal answer. Let the old buzzard stew in her own juice. The day seemed suddenly sunnier and she began to plot to get rid of the old lady.

She was foiled.

Paddy, in his male innocence, came into the kitchen, full of pride at their coming excursion.

"Ah, Mrs. McBride, is it? Isn't this the perfect day, entirely? I've borrowed me cousin Roddy's car, him that's a mechanic down at the garage, and we're off to the countryside! We're after having a little picnic and a little sightseeing." Then with what Moraine thought was insufferable arrogance, "I'm taking me whole family out on one of our regular outings."

Moraine could have throttled him. Regular, indeed! You would have thought him the laird of the manor. He had blown her bubble of satisfaction.

Then there was worse to come.

"Why don't ye come along with us?" Paddy's kindness did not strike Moraine as a kindness to her. She would have to round up more food, more cutlery, and probably another flannel for that old bat in case she spilled her tea. She would have to spend her whole day being polite and noncommittal. Moraine drew in a breath. Surely, Mrs. McBride wouldn't accept the invitation.

Sure and she did. Lonely as she was, she was almost pitiful in her pleasure at being included in anything. And, even though she would not cease to peer around at the arrangements and at how the children were disciplined, or not, she had been mollified, temporarily, by kind attention. Something could be learned from Paddy's approach, but it was nothing that Moraine wanted to learn.

The day was a roaring success, unless one takes into account the fact that Maeve and Kevin were in competition all day.

For a while, it looked as though there was no getting away from the tour buses. As they tried to negotiate the turn into a curvy narrow lane, they had to squeeze past a pride of tourists who were stopped for a spot of tea.

"If I never see another emerald green slope covered with snowy white sheep, it will be too soon," complained one woman. "Where's some action? Maybe a fight or something like that." Moraine was incensed. Her Ireland! To think that it could be a goldfish bowl for sensation seekers.

In the car, the battle was over territory for the children. Mrs. McBride was made the boundary over which one did not reach a grimy hand. At first, they were a little cowed by Mrs. McBride's presence but it did not last and when a glen had been selected for the picnic, off they went chasing around the rocks, hiding and then jumping out at each other.

When they found a particularly magic place in the rocks, they would immediately set up housekeeping, lining out households with smaller moveable stones. They found sticks that could be used as forks and from the few trees they found leaves for pretend placemats.

"You're going to disturb the little people," protested Moraine. "You know they live in all the little cave places in the rocks. And they play tricks on those who invade their privacy. They'll turn you into a goat with a long beard." It was her hope to scare the bejesus out of them and give herself a little rest.

This got Moraine about a fifteen-minute reprieve from mayhem before they were off again. Her storytelling had promised them that if they caught a leprechaun he would be forced to tell them where the pot of gold was buried. It is hard to say whether they treasured the thoughts of the gold more than the thought of a little, testy green-hatted man of their very own. She told them about the time her great-great-grandfather had come upon one of the little ones.

"Most of the old men of that day were peat cutters, but my Grandda was a woodcutter. Of course, there wasn't a lot of wood to be cut, so he was wandering in the woods looking for a decent tree, when he stumbled over a toadstool. He picked himself up, dusted himself off, and would have gone on with his search, when he heard such a cursing and ranting that he looked all around for the source. Finally, he looked down and there, pulled up to his full furious height, stood a little man. He was standing beside the wreckage of a lovely little house that, before the big foot had smashed it, had sported a balcony, a rocking chair, and a splendid outhouse with a half moon cut in its door.

The poor old woodcutter apologized over and over for his terrible mistake, but the leprechaun would have none of it. Finally, my Grandda pulled a flask from his pocket and offered him a snort if he would calm down. Somewhat mollified, the heroically named Finn agreed and they sat down to talk. After a while, the little man kindly agreed to permit Grandpa to build him a whole new house with his considerable skills as a woodworker. So Finn made himself a little shelter and waited. When the house was completed of the finest woods, it looked fit for the High King's Tara. Finn had to admit, 'though with some grumpiness, that the house was worthy of some reward. He climbed up on a tree stump and commanded Grandda to kneel and he pulled out his sword - which looked like a long darning needle - and knighted him on the spot. From that day, Grandda had the power always to get revenge on anyone who wronged him. And I inherited that power," finished Moraine, threateningly. Delighted, the children ran off.

The picnic food itself was slightly less than a success. Mrs. McBride had to make some remark about the leaky condition of the basket.

"Dear, I would have been pleased to have lent you my almost new basket had I realized you would have to deal with this -" her voice trailed off.

Besides that, the cake was more gritty than spongy even before Kevin dropped his in the dirt and Maeve spilled her hot tea down her front and howled and howled and then Paddy fell asleep under a tree and both children had to go to the bathroom where there was none and Moraine had to search out some old newspaper from the car to use for toilet paper.

Mrs. McBride looked on with relish. They had not thought to bring a chair for her so Paddy had found a flat rock in some shade and had put his own coat on it for softness. She was not entirely satisfied by this arrangement although Paddy's charm smoothed her feathers to some degree. Notwithstanding all of the kindness shown her, she was making a little mental note of errant behavior with which she would regale the rest of the neighbors as soon as possible. She really felt that Moraine was a mite too indulgent with those little hellions. In truth, Moraine really just wanted to avoid them as much as possible.

Moraine was getting a little heavy in her pregnancy, a little awkward in her gettings up and sittings down, but no place had been provided for her by her ever-so solicitous husband so she chose to walk around and down to one of the ancient dolmens. The rocks kept her hidden, she thought, but Maeve and Kevin spotted her and came running for a story. They had been playing King of the Mountain, each trying to get to be on the highest rock, which is how they were able to find her.

The story that came to Moraine's mind was that of the shrieking stones. There are purported to be many such in the Celtic tradition. Moraine told them of the *Lia Fail*, which was discovered by the great king of Tara, Conn, when he stepped on it and it shrieked. It became known as the Seat Perilous that screamed thereafter if an unworthy knight sat upon it. The stone did not stay in one place, and would show up in any village where there was to be a coronation. It was the only way a new king could be recognized, but woe betides a false king or queen who tried to claim the crown. The shrieking would expose their calumny.

A grand new game evolved for the little ones! They pushed each other from every stone in the glen. They were bruised and bleeding and very satisfied. They searched endlessly for the *Lia Fail* and they did their own shrieking.

The trip home was surprisingly pleasant. Maeve and Kevin fell asleep curled around each other like puppies, all warm and sticky. Even Mrs. McBride, who was stuck in the back seat with them, fell asleep with her mouth open, a bit of drool and a bit of a snore proving her condition. Her hat slipped forward making her look even more like a bird of prey.

Moraine and Paddy said little on the way back. There was a comfortable silence between the two who had been through years together. At first Moraine fished around in her head for some interesting tidbits to relate to Paddy, but there seemed nothing new in her life to relate. They both subsided into a quiet rhythm of bumping along in the old car. Her day in the warming sun had caused considerable thaw in her glacier. She nodded contentedly, feeling almost slushy.

When they stopped for an ice cream at the same shop where there was a tour bus, they were not too surprised to find the proprietor's three-year old daughter sitting on the ice cream freezer lid and his eight year old son standing guard over the postcard rack. The little children had been posted to avoid the sticky fingers of the busload of tourists. Moraine thought, *'Oh, those Americans!'* Then she thought of her cousin, Bridey, now living in Boston.

CHAPTER 15

Green, Gold, and Rosy Colours

Dear Bridey:

We've been on a picnic! We actually got away from home while the children were in school and only the grands were with us. We wrote notes to their teachers saying they had sore throats.

Can you believe that idiot of a husband of mine invited Mrs. McBride? It worked out all right, but I could really have done without waiting on her. I am so pregnant I can hardly bend over and even if I could, surely would never straighten up again.

Maeve kept wanting to climb up in my arms, sticky and gooey though she was. I was still so put out that Patrick had brought Mrs. McB. and then dumped her on me to entertain that I really didn't want to be bothered. Later, when I was home in bed, I started questioning myself. The little one might not ask again; she might grow all the way up without that hug. I have missed a chance to come closer to her and she to me.

My neighbor, Ellen, who has the child with Downs Syndrome, has been able to place her for a week of respite care. There her daughter will be kept and loved and treated as a guest, and Ellen herself will have been given space. The nun, a Sister of Mercy who keeps the facility, can have only two boys or two girls at the same time. It is her life's calling and she keeps the place just like any home except for such as the

93

electric lift in the bedroom. She has volunteers; one man sleeps on a pallet when there are boys to be attended. The occasional novice helps bathe and toilet the children. Today is only the second day, and already Ellen's Parkinson's disease seems to be eased. She even has on earrings and has a dinner date with her husband though she sorely misses her child. As Granny always said and I do mean always - "The Lord niver gives ye more than ye can bear." How did Granny know?

I've been dreaming of buying a new dress. Do you not think that a nice rosy color would be better for me? I've worn 'practical' until I shall just have a fit if anything practical, black or brown, appears before me. It must be the cloudy weather. If I promise my God and all of the saints that I won't spill tea down the front, do you think I could have something that would brighten the day? What are they wearing in Boston?

Best to you,
M.

Dear Moraine:

You seem to have such an exotic view of our life in Boston. I never told you the half of it. Emigration is a much bigger step than most young couples believe. We should have found out more and maybe moved to another area. As it happened, we settled in South Boston, which is the centre of Irish life here. It felt just like Belfast. Our men have the most awful job finding employment. It is not quite so frightening as it was when all of the men had to work in mills or build railroads or anything dangerous or menial. David is finishing up a plumber's license as he works as an apprentice. That has taken a long time but will make our lives a little more secure. The pay is splendid.

The main trouble was that we were still isolated from Americans, even after out citizenship was assured. We wanted so badly to become citizens, but wanted to hang on to our Irishness. No problem. We were expected to wear green every day of the week and say 'Faith and Bejasus' about once an hour on the hour.

The pubs all have Irish names: O'Reilly's, Flanagan's, Murphy's, and, most incredible of all: *Tira naGog*, Land of Eternal Youth, where, incidentally, there is usually a pretty good band. The storefronts in many cases are boarded up, the houses are expensive and becoming more so. I am afraid that we have become gentrified nowadays. Because of the waterfront, new condos are going up all over. It is quite the stylish thing to be living on the water and a lovely sight it is, too, in spite of the commercial parts like cranes and front loaders.

Of course, this is not pleasant for those who have lived here for years; have talked over the back fence while hanging out the wash, and who do not suffer fools lightly, but they had to make sure any fools suffered were their own fools. 'Stick-with-your-own-kind' fools.

So, we are the same people simply transplanted, lock, stock, and barrel, like an uncut shepherd's pie, moved from one spot on the counter to another. Our women are still the controllers of the family, the examples for our children, the carriers of tradition. In a way, this is changing because women need to work out of home, so the young ones are not glued to us day and night.

It has been said that if the Irish had no drink, they would control the earth. We'll be a while finding out about that, certainly in my household.

How did I ever get on such a roll? This amounts to me talking to myself like an *eejit* whether or not you want to read it. You are my confessor, my therapist.

Best to you. Start saving your money to come to this golden-street country.

Love, Bridey

CHAPTER 16

Hope and Promises

Dinner had never been a very shining affair at the O'Shea's, what with Moraine's slipshod cookery and Patrick's frequent absences or domination when he was there.

Now with her pregnancy, the whole domestic situation had slid into a scene where grease and pure dirt prevailed. She tried to keep things presentable but often felt apathetic about the whole thing. She was sometimes tired just from doing nothing while chores mounted. She had not learned to recognize depression, especially in herself.

Moraine was a stockily built woman. Now being in the family way increased her bulk and also her discontented mien. The joy of a new baby was pretty much dimmed in her memory, though Father Reilly was full of telling her how clever and saintly she was to be bringing forth a new life.

What a phony old fool he seemed to her! Babies cost money and messed diapers and threw up. She never planned to help repopulate the earth just for another war.

While carrying her fourth child (not counting two miscarriages), her skin didn't glow, her blue eyes were dim, and her brown hair was stringy. Yet, at the time we begin to recount, she was almost forty-five years old. She should have been able to stop having babies about now.

"Ah, you should be seeing the grand new

bar maid at the C&C," Patrick would report. "Red hair and green eyes and big tits and ready to laugh with a fellow."

Then Moraine felt like a failure. She was too ashamed to tell Father Reilly that she just wanted to be held in Patrick's arms, to feel wanted and beautiful - and what would he be thinking of her and what could she say to the Blessed Virgin above, as pure as snow? There was a need for an intimacy of which sex was only a small part, a time to share deep thoughts or even silly ones. There was still a feeling that she needed a man in the center of her life. Where did that come from? Not Queen Maeve. Not as a center with no controls at all. She wanted to punish both Patrick and the redheaded one. Her Irish possessiveness flared.

Moraine's mother had never told her that the flame could subside. Maybe she assumed that, just by being a woman, Moraine would know that. Maybe she never knew there could be a flame.

Moraine could see her mother now, with her reddened hands, a dishtowel around her thickened middle and that everlasting black shawl that was always there to pull over her hair. In fact, Moraine had that very shawl way back in her bureau and she purely hated it. Now, she began to see that her Mother had thought that there was no more to life, that her Mother had been deeply lonely, trapped in the 'joy' of constant childbirth, urged on by wifely 'duty' and the almost certain knowledge that Moraine would follow in her footsteps, as all women were expected to follow the footsteps of their mothers.

Even years later, Moraine would not regret

the terms of her marriage to Patrick, though not every day. She had begun at that very moment of wedding promises and hope to pull the shawl-like glacier from her head and her life.

The O'Sheas were not 'bog Irish', but they were not comfortably set. Moraine aspired to more and took great pains to listen carefully to the telly, trying to imitate the posh accents of the commentators, British or Irish. This maddened Patrick who considered all such attempts as phony, trying to be above herself. Only in times of stress (and there was plenty of that, what with Himself often not present and the children growing up), did she revert to her old way of speaking. "Whisht, all of you! Is it daft you are trying to send me?" in a voice raised to a screech. Moraine tried very hard to be the calm nurturing person she thought she had to be. The real Moraine kept cropping up.

One of the reasons she thought her marriage was all right, despite Paddy's constant drinking, absences, and tendency to strike her, was the other marriages she had knowledge of were just the same. She had been so naive when she married.

The other reason was a more insidious one. Moraine was an enabler. The hallmark of an enabler is the justification of the partner's activities and she had this quality in embarrassing quantity.

"Shush, shush, now. Your Da has had a hard day at the plant and is all worn out and stressed entirely." This she said to the children.

The children saw right through her words, knowing that their Da had been out drinking and this in spite of the fact that he did as little work as possible. All assembly line jobs were nobrainers. Then he had his pack of cronies to coffee break with, smoke break with, goof off with and then to meet in the neighborhood pub while the family waited for him at home. His life looked like Paradise to Moraine.

The term 'stress' came into flower at just the right time for use by Moraine. Much could be blamed on 'stress' but she forgot to apply it to herself, only to justify Paddy. On the other hand, he did bring the bulk of his salary home and that always justifies a lot. This was a fact of which he never failed to remind her. He was the breadwinner. He brought home the bacon. He was the master of his castle. It left her beholden. She could not ask for more without being reminded how much she owed from his point of view.

As Moraine thought of these things, anger began to build inside her. The heat of her fury began the day when he proclaimed himself master one time too many. It began to melt the glaciers of her learned old traditions, drip by drip. Drip went the old mind-sets, the habit of blind, almost stoic, acceptance of any conditions. She could have made choices! She realized that she could have had a better life single and working although at that point she would think about the emptiness of life without Patrick. She could have gone to America if she had only had the nerve. She once had a good job teaching pre-school, a job not available to a married woman.

Her mind was wasting! Her spirit was shrinking!

Her mother had told her, as her mother before had told her, to be accepting, unquestioning and, above all, to be humble. Moraine must have been pretty fertile ground for this planting but as she began to look around, she saw that there were many women who had broken the mold and now each had started to think for herself. She had been too gullible, never really being mindful of herself. Even her prayers had become automatic, with no thought of content.

'Humble, shit.' thought Moraine. *'Wait until this baby is born. I'm going to be a brand new person, my very own self.'*

The glacier had only begun to melt and such changes take time.

CHAPTER 17

Discord

The days that followed during 1978 were bittersweet. Moraine poured her energies into her grandchildren as she grew large with her own fourth child.

The children missed their mother but most of all they missed their freedom. Maeve and Kevin had spent their lives mostly to themselves and indeed had come to rely on each other in the absence of a stronger figure. They were incensed at this intrusion into their partnership. At any request, one might have thought that she was ordering them to scrub the kitchen and cook the dinner.

Moraine insisted on nightly baths and finally chased them down, flinging them each into the tub with their clothes on. All three of them were soaked. Moraine failed to see the humour of it all in spite of Patrick's roaring laughter.

They tried to resist her directions with every weapon known to children, especially whining and rebelling. On the one hand, they had to admit deep inside their little selves that they loved the relatively orderly household that Moraine tried to maintain. There was always at least one clean outfit. Well, maybe the socks did not always match. On the other hand, they resented the fact that she had to 'know everything'.

"What are you whispering about? What are you chewing? Where did you get that?" She was so nosy. Mom had ignored stuff like that but

Gamma was into every little thing. Her experience with Kathleen's rebellion had taught her to ask questions even at the risk of unpopularity.

Moraine had looked forward to the morning walks with them on the way to school. She had expected to point out the flowers, the old toads, a tortoise, any worthy object as she had read someplace in a magazine that mothers should. She had this sunny picture of herself but it turned out that she was only being kidded. It was raining most of the time, leaving everything in sight as drab as Moraine felt. Only her flowers flourished from the rain and cheered her with their beauty. She drew courage and then resolve from the colors of her flowers. She let the icy feeling around her melt a bit.

Now she only wanted to expediently get the children to school and get back to her own chores. Her image of the happy trio had also been shattered by their attitudes. She felt compelled to smack the behind of the complainer when she had heard enough. One would lag and the other would run ahead. The children began to suspect that maybe she didn't really care as much about them as she said. Even though she loved them for sure, the one thing she could not abide was a whiner.

"I hate this mess for lunch. Why can't I buy lunch at school like the other kids? I hate you too."

"Gamma, he's hitting on me. He's walking on my side of the sidewalk."

"Why can't we walk to school without you tagging along? The other kids can." They could not because of the fear that floated in wisps like a miasma, like the fog that had covered the potato fields during the Famine, killing everything. The other parents were undergoing the same sort of battle of trying to protect. They were walking their kids to school no matter what Kevin and Maeve said about the kids allowed to walk to school alone.

In July when the Orangemen came parading past (on their way to go through the Catholic area) Moraine watched the children gathering stones from the garden edge to throw at the marchers. Instead of grabbing the children by their ears, she wanted to join them in the street. She turned her back with a half smile and busied herself dusting the living room tables. Dusting was not a natural activity for her and she hated it.

She wrote her name on the table and then dusted it away. She broke another ceramic doodad of babies with sand buckets. Then she peeped out around the curtain to see what was happening.

It was said that the fierce beating of the big drums called *lambegs* could beat out the Pope. The sound of the *lambeg* beating out the rhythm reverberated on the piano keys of her spine.

Had she been out there she would have been able to inhale enough alcohol fumes to make her giddy. Patrick was out today, with his lodge, with his derby hat, his bumbershoot, his orange-fringed scarf. She had ironed it for him and fixed his thermos of tea that he later laced to a good degree. She had done her wifely duty though she really hated the devotion of the Protestants.

That evening, when the marchers were to return, Moraine hurried down to the Main Street expecting to join a crowd welcoming them back. She had hoped to be hidden in the midst of a lot of people. The street was empty, inhabited only by police and soldiers patrolling. There had been signs on all the windows and telegraph poles earlier asking for people to come boo the returnees. Someone had torn them all down. The result was that everyone was laying low wondering if they were being watched …… and by whom?

Later that night, when groups of youths had gathered, hoping to throw things at the marchers, the marchers came in by a different route and returned to the Hall in a secret way. At least they did not try to return to cause more conflict. That was a small peace sign. It seemed that everyone was tired of the mess that they didn't know how to stop.

Yet Moraine was still a Catholic, once a Catholic always a Catholic. How could she put up with Patrick and his flaunting the 'other side'? Today, while thinking about the children out in the street, Moraine managed to break a couple more ceramic doodads, bringing her score up to six in the past month. She was delighted. She should have thrown them. It bothered her, though, that she had ignored the children's stone throwing. Maybe, just maybe, this is another piece of our lack of forgiveness. But then, she thought, those marchers did not have to come right down the Catholic streets. What really bothered her mind was the fact that she had taken no stand, no positive vote. She had let the children think she felt one way and she wasn't even sure of that. Children know truth.

In spite of their alliance when Maeve and Kevin were alone together, when they were with Moraine all kinds of spite surfaced between the siblings. Moraine's answer to disputes was usually, "Push him back."

She demanded and didn't get complete *smacht*, obedience. She had started out trying to settle each spat separately but the nearer the time came for her own baby, the shorter her patience became. Rearing grandchildren was not what she had expected, nor was this baby anything she had bargained for. And her back hurt.

It seemed that Patrick was only there for the good times and then he disappeared whenever conflict arose. The children soon learned to use him for shelter from the storm.

The cats that Patrick disliked so much were peacemakers, too. Bel and Sam (she knew he should be called *sow* in the Irish manner but it didn't ring true) were stray cats, Beltane a tiger cat and Samhain a calico. They had come mewing as babies on a rainy, windy day and lonely Moraine could not turn them away. Sometimes she wished she had, but now with the grandchildren, they had a new use, a cuddling use. When either of the children was in need of a hug, a soft cat would be sought out. Of course, the cats decided whether it suited them or not but it was a sure cure for whatever ailed the kids.

Then Angela was born. She was a lovely slip of a baby who cooed and reached with her tiny fists and sucked her thumb. One day she managed to grip her wee toes.

Moraine had expected Maeve and Kevin to be equally joyful and very helpful. She reckoned that she could push some of the care off on them but she found that she was wrong.

Babies were really not Moraine's thing. Moraine had never really wanted children, never even had that dream as a young girl. She had played with grown up dolls like Barbies, not baby dolls.

Pregnancy was something to be expected. It was the curse of womankind. It was the normal course of events.

So her babies had come, Kathleen, Brandon, Liam, one after the other, and now Angela.

In the generations before her, women often had fifteen babies, sometimes even more, until they died in childbirth, worn out. 'Nerves' prevailed. Their husbands were in control and the Church backed them up in their desires. She had never been able to convince Patrick that she wanted no more after Kathleen.

Even the prettiest of infants wet their pants, and worse, were either hungry or spitting food. The grandchildren would have none of it. Now they had something on which to focus their pouting.

"You spend more time with that kid than with us."

"You're not the boss of me."

"You're not my real mother."

They dreamed of going 'home' and it was something they never overcame. They were so deeply hurt that they could not stay with their parents. Home was a fantasy that had never happened. The awful cellar room they had come from became transferred, in their dreams, to a house with a garden and a gate and a separate bedroom with pink in one and blue in the other. Toys were endless, and the bicycles, oh, the bicycles were such gleaming beauties.

Moraine's mental picture of herself at this point was hilarious and pure fantasy. She saw herself as a matron with all of her children and grandchildren gathered about her knees for succour and comfort. Jesus Christ! How in the world could she justify being both lukewarm about children and being such a mother?

It's funny the way one develops a self-image. Moraine looked upon herself as the Saviouress of the children, the great Earth Mother who sheltered all and sundry. But it was Patrick, with his face of an aging leprechaun on a tall, lean body, who made the days fun and bearable and in the end, it was his influence that kept the little ones in the nest.

Patrick loved to sing and he loved little children - no matter what he thought of them when discipline was required as they matured.

Seated around the kitchen table, he taught them every Irish song he had ever heard including a few that Moraine had thought were a mite raunchy from his hurling days.

If the three of them heard Moraine approaching, they immediately launched into 'When Irish Eyes Are Smiling' and pretended it was just for her.

Maeve and Kevin particularly liked the sound of 'Mrs. Murphy's Chowder' because it gave them license to sing with ascending volume, finishing either with a splendid crescendo or accompanied by the sudden bang of a pan as Moraine signified that she had heard enough.

The children looked forward to Patrick's time at home, all the more wonderful because it was at a minimum, what with himself going out with the boys.

Moraine felt herself growing frustrated with the whole situation. Didn't she feed the little monsters, clean them, tell them stories, and love them? How come Patrick got all of the attention that she wanted and needed so badly?

She had to admit that the times in the kitchen, with the cabbage boiling, the house all warm, and the music ringing out, were the best of all times.

The discordant notes were her singing, which she insisted on doing despite requests that she stop. There were also the discords of her feelings of jealousy. She desired to have more of the grandchildren's attention, or more to the point, more of Patrick's attention. Possessiveness is an Irish trait. A curse.

Her nightmares still came on, but rarely was Pat there to comfort her. He didn't even realize she needed him.

The kitchen was the only room that got any use. When any of them came in the house, the first act was to throw any books, coats, or packages on the table which meant that all had to be cleared when mealtime came.

By tradition, the living room was saved for wakes and the big Bible that recorded each birth and death. Otherwise, the Bible didn't get much use. The Irish prefer to relate stories of their own lives, tales of the fairies and supernatural occurrences, and the maybe the more lurid of the Bible stories that they had heard since childhood. Dust grew thick on the old Book that had been in Patrick's family for generations though rarely opened by the more recent ones.

When Moraine told stories at bedtime, they were almost always myths of revenge and control. Those were two things she, like her mother, lacked and dreamed of acquiring in her life. The stories seemed to be a release for her feelings of being unappreciated.

When she told them the wondrous tales of Finn MacCool, she always emphasized the fact that Finn thought his mother had exhorted him to revenge. When Moraine found out later that he thought his Mother had supposedly abandoned him on a bog, she decided not to tell the children about that. It was a little too close to the mark.

She never quite got around to telling the whole tale about the way his wife, Grania, nagged him and pushed him around until the day he died. She just explained the parts about Grania's control of his household. Perhaps she should have told them the whole thing. It is from stories that we learn truths about ourselves. Either way the story was told, there was something very, very strong about his mother. Moraine preferred tales of strength and revenge, successful revenge.

She was not sure for what she wanted revenge. She wanted to get back at Patrick for his neglect, back at her Mother for telling her to be 'nice', back at her school for teaching the old female stereotypes. She wanted to get back at somebody for the Troubles. She had no way of gaining the control she wanted. As in the search for the Grail, she didn't know where to look and she didn't realize how close it really was.

Time after time, mostly night after night, while she waited for Patrick to come home for dinner, she would anxiously prepare herself, heart wide open and vulnerable, so full of love and hope. She would put candles on the table, have a cold Guinness ready for him, all of that romantic stuff. Maybe she failed to smile and laugh for him. Time after time, night after night,

Paddy missed dinner and the whole evening.

"What do you care where I've been? I don't have to hang around you all the time. I don't even need to give you an excuse. You have nothing to do but sit around here all day while I'm hard at work. And don't you be checking the pub to find out where I am."

Her face had been duly slapped. She could feel the phantom red finger marks on her cheek. The aperture to her love-awaiting heart slammed shut and she began to recede. Once again, an icy layer built up around her.

In her callow youth, only so few years ago, she had expected him to enjoy all of his time with her. Now she felt he found her worthless and boring. She never quite overcame this assessment of herself, even when Patrick later pleaded with her, "I was only joking."

Expectations - the ruination of us all.

CHAPTER 18

Gift to the Church

Moraine luckily had only four children (luck had little to do with this). She was not one to be really sentimental about a wee one. She did try to give the impression that she was a natural mother, but sometimes it came out all wrong. She did, after a while, realize that a lot of her chronic tiredness was a symptom of depression. She really needed to get out of the house, so her forays with education saved her life. It was a lot easier for the children to put up with her, as well.

Liam was Number Two child and Number One son. Patrick was disgustingly proud of the fact that he now had a boy. The boys at the Pub tolerated a lot of bragging until they deemed enough time had been spent on discussing Liam and told Patrick to shut down. He wasn't the only father in the world with a son.

When Liam was a baby he was chubby and adorable. If Moraine had not been so busy being angry with Patrick for his perceived wrongdoings, she would have enjoyed Liam much more.

As a wee lad, Liam was round-faced, freckled like an overripe banana, and possessed of the runniest nose in Northern Ireland. Kathleen flatly refused to have anything to do with her baby brother because of his 'grosbeak', as she dubbed it, and Liam toddled after her futilely hoping for attention.

As for his mother, she hated his whining and told him over and over "boys don't cry" which only made him cry more. Her words did not give him courage, only the courage to cover up his lack of courage. He became a bully to cover up his weakness. He took toys away from other children and refused to share his. In all of this, he was forgiven by Moraine who thought he was being 'all boy'. Besides, it was right much trouble to correct him consistently.

As he grew and started to school, Liam became even worse. His Mother exhorted him to make friends, to do whatever (within limits) would make the other boys like him. Liam had no wish to be compliant, so any attempt to fit in went against his grain. If he did try to be one of the gang, his Mother was furious with him, and Paddy even more so. He was damned if he did and damned if he didn't.

Poor Liam didn't know which way to turn. Everything he did was wrong according to someone. His bullying grew worse, and his need for a place to belong grew greater. Once he outgrew his cute baby years, he grew into a terribly awkward stage complete with pimples, and girls would have nothing to do with him. He read every bit of pulp literature he could find in an effort to find the magic spell that would bring them to him in droves, but the only result (not a bad thing) was that his reading improved mightily while his thoughts merely took a dive. He sent for every free muscle-builder, skin-clearing book he could find and especially the ones that promised he would earn a million dollars by the time he was twenty-one.

As ancient tradition demanded, the Number One Son became a man of The Church, but Liam did not become a 'religious' for the traditional reasons. He had no real calling, but thought this would be a life of glamour and acceptance, an escape from daily events.

He had no wish or understanding about giving back with his heart. That would have to come later.

This turn of events was a matter of great pride to his Da. Patrick would not only have an advocate with the Father (Himself Above and Father Reilly right here in the parish), but it cost him nothing and had accrued much merit for him amongst his fellow employees and pub pals. This, in the face of the fact that Patrick was a backsliding Protestant who only showed his other face when there was a great march of the boys and a great singing and drinking. He could enjoy either side and talk politics with the best of them. Lately, though, he had felt a need to find some serious ground in the politics that he could support. He was no fool, so he started listening a little more carefully and reading the newspaper each day for information. He no longer believed everything he heard in the Pub.

Moraine and Kathleen were going places but he had no idea of where. He did not understand that they, too, were searching for some truths, and their dinner conversation did not improve. Nor did the dinner. They all disagreed on everything.

It was a great and glorious moment altogether when Liam announced his decision of religious vocation at dinner. Sadly, everyone had forgotten to ask the blessing before diving into the usual boiled stuff. Maybe it could have helped to say a little something to the Creator. Moraine was not noted for her cuisine but a hungry person could be filled and it was always ready on time. More than an ordinary blessing would be needed for the repast for which no one was truly thankful. Her Granny would have recounted the days when there was nothing to eat. She was gone now and the current generation had no clue about the reality of the hardships that they had heard about so often.

With his mouth full, Liam announced that he had been praying with Father Reilly for about a week and he figured he might as well get on with it. He was instantly remorseful that he had not made a more dramatic announcement with suspense and flair. Too late. He had even memorized a formal announcement speech that stayed forever unheard.

"Hooray!" burst his Da.

"Are you daft?" asked Kathleen of both Liam and her father. By her very attitude, Kathleen gave the impression of chewing gum at all times. Her so-called taste ran to very brief skirts and she would have sported fishnet hose but Moraine had forbidden her to enter the house in them - this in a very convincingly loud voice and with accompanying bodily threats. Naturally, an extra large bag was required for the hosiery and extra eye makeup to be smuggled abroad and donned in the bus station rest room. There was hardly room in the bag for schoolbooks.

"You are going to be poor and celibate? In a pig's eye. How come you are planning to become so pure? Are you going to be like Jesus Christ? You've never even made your bed or washed a dish, for God's sake."

"Kathleen, watch your mouth," roared Patrick. As head of the family, he was full of pride at the very thought that his boy would be an advocate to the Father Above even though he himself did not hold with such papist carryings-on. What he could tell the boys down at the Cork and Clover! How he would strut! "My son, Father Liam"! How that rolled off the tongue. Liam might even take on a name that had a more sacred sound, Peter or Pius maybe. Something Italian might be a validating touch.

Down in some unconfessed cranny of Patrick's mind there lurked a deep relief that he would no longer have to pay for food or clothing for Liam once he was properly launched. Such thoughts were kept out of sight so that the great 'calling' would not be negated.

Although his father seemed proud enough, Liam was a bit disappointed in the general response to his earth-shaking announcement. He had envisioned himself as handsomely holy in roman collar and black robe with huge crucifix dangling. He wasn't too sure about this tonsure thing. His Father already had a bald spot that he was trying to conceal. He could see himself hearing confessions and patting little girls and boys on their heads - all eyes would be on him admiringly as he performed the mysteries at the altar. Had his family no respect for the Cloth?

Moraine looked at him with the lovingly blind eyes of a mother. *'Wouldn't he be the lovely priest, though? And to think he is our gift to the Holy Church'*. The term 'pompous ass' never entered her still glacial mind. Nor did it occur to her to consider her own doubts about the Church.

CHAPTER 19

Fathers

Liam found his novitiate less glamorous than he had anticipated. He had learned early on to avoid any kind of labor. Like Moraine's brothers, he had been pretty much allowed to go free while the women and girls did that kind of thing. Now he learned, to his distress, that when he shirked a job, the whole brotherhood suffered and had to do it for him. That probably would not have bothered him, except that he still wanted to feel liked and, although the brothers never showed disapproval by facial expression or even manner, the distance was maintained. He was not part of their laughing and singing because they did all that while doing their chores. He managed all of the studies very easily, thanks to a surprise strain of intelligence that had been previously hidden from sight.

At first, he floated about barely doing his duties, attending to his studies and learning to appear obedient. Without Liam realizing what was happening, the abbot kept an eye on him and gained a pretty good handle on his inner workings. He thought maybe Liam could use an awakening. He asked Liam to accompany him on his monthly visit to the slums, the poorest of the poor, and help to carry some of the food and clothing to those in need. Liam was quite proud to have been selected, never dreaming that Father Andrew had an ulterior motive. For a few days, he strode about, feeling quite priestly, feeling superior to the brothers who had not been chosen. It didn't last. The awakening was on its way.

Off they went, laden with all of the food and clothing they had been able to scrape together in a month's begging and scavenging. The old rattly car, which they were lucky to have in the first place, threatened to die each time they slowed down. The blockades, searchings, and questionings by the new ever-changing soldiers (although Father Reilly came through about once every month and was a familiar figure) were exhausting and demeaning. The most tiring part was the constant humiliation, because Catholics were especially suspected of some unnamed intention.

Liam was to learn on that first day what 'poor' means. It means that one who is homeless cannot return home for a warm, clean bed and a hot meal. There is no place to return to, no recourse. When they set up their mini-soup kitchen and started ladling out cabbage soup and handing out chunks of bread, he found the need to keep his mind away from the actual pain, both physical and emotional, on the faces he saw. It proved to be impossible. He blamed his wet eyes on the steam from the pot.

"How do you stand going there all the time?" he asked Father Andrew. Father Andrew mopped the top of his old head, bald without assistance of tonsure. "My son, I have learned two things." he replied. "One is to suffer with each soul that we meet and try to fulfill the mandate from Christ to feed and clothe the poor. The other thing is that when we return to the monastery, I must pray to be able to let the pain of the day rush out of my soul like the brown waters from a broken dam until the waters are clear and finally still. This image stays with me and cures, but in reality I have unburdened once more onto Jesus."

Then he added, "Next time, we are going to the Methodist mission downtown to work with the people in need there." He gave a little sideways glance at Liam to see how this bombshell would affect him. He was not disappointed.

"Do you mean to stand there and tell me that we are going to work side by side with those dirty Protestants? We'll be in harm's way even going into that district. What are you telling me?" Liam asked.

These Protestants were people like Liam's own father.

"Listen, Liam. Listen good. They are people, human. Their needs are the same as ours. If we don't help, who will? And where, then, is peace? One of my dearest friends is Reverend Goodacre who sometimes meets us to help with the supplies. You will really like him. He knows a million jokes."

Liam was stunned. Just people! Jokes! He had thought that revenge on them for their nationalist leanings would have been the only way. Hadn't his mother led him to believe that, even while she generously said that Da was entitled to his beliefs? That didn't even make any sense then, come to think on it. Now, he would have to swallow his pride and work beside 'them'.

When Liam and Father Andrew got through all of the barriers and entered the slum area, Liam expected no surprises. He thought he was fairly familiar with 'poor' areas, having frequently walked past the peeling paint houses and the broken gatehouses. What he wasn't aware of was a place such as he and Father Andrew visited next. The street was well paved; the turn into the narrow side street looked well kept. Then it all changed. It was a rabbit warren. Within these walls, slated for the wrecker's ball to make room for development, were rats, bugs, garbage, backed up sewers and at least ten people to a room.

The first body blow was the never-to-be-forgotten smell of mingled garbage and sewage. A resident appeared at a doorway, peering around furtively. Then he disappeared to spread the word that food was on the way. Another head appeared. That person hurried away, fearing to be found by the *garda*. Furtive may have been the key word in any description of the area. Even the garbage cans looked furtive as though they were ashamed to be there and were afraid of being caught. Fear walked the street.

Were these Republicans or Unionists? Who might set a bomb here? Never mind. They were all poor and their makeshift homes were now going to be destroyed. Where could they go? They were promised new places currently being poorly built and being decried by those who would be neighbors. "Not in my neighborhood" was the key phrase.

At the end of the block, at the top of one steep unpainted and leaning stairway, a startling vision appeared. Huge shoes splayed open to contain the swollen feet and ankles of an old, old lady. Old in years? Or old in poverty? Liam thought she looked like what his Ma would have called, 'rode hard and put up wet'.

In spite of the day's heat, the dear lady was

an outstanding example of the style of the day, known to the modish, as 'layering'. She was garbed in a turtleneck, sweatshirt, a jumper, an insulated vest, and a shawl - and that was just the visible part.

"Do come in, gentlemen, and have a wee cup o'tea with me."

Father Andrew let out a big sigh, inch by inch so that his hostess wouldn't notice. He had hoped to avoid her this day. Oh, those steps!

"Ah, that's darlin' of ye, Lizzie. We have a small parcel for ye."

The stairs were no easy challenge for a man with Father Andrew's arthritic knees. Praying on his knees had become a literal sacrifice to his God and getting back up was worse. Liam carried the bags of groceries and clothes so Father Andrew could pull himself up by the rickety handrail.

Lizzie had put the kettle to boil by the time they got to the landing and peered into her castle. She had a small gas plate to heat the water that had come from a rusty spigot. Her feminine urge to nest and decorate was overshadowed by her pile of rags used to keep warm and afford some protection from strange creepy, crawly critters that traveled while she slept. The tea table was leveled by a book under one leg, *Knowledge of the World - Complete With Full Color Illustrations.* Lizzie had about as much knowledge of the world as she could take. It satisfied her greatly to see the mighty tome doing some good.

Already set on the table were one mug from Bewley's, one chipped and flowered cup, and one blue cup with no handle. There was one teaspoon and one teabag, some packets of sugar from a nearby restaurant and a tin of questionable milk. One must not overlook the jar in the center with fragrant red honeysuckle from the nearby park (who could believe a park could be nearby this warren?). The fragrance was welcoming and a good reminder that Lizzie was aware of the 'finer things'

With exaggerated dignity, Lizzie invited them to sit on one kitchen chair and one wooden crate while she eased her bulk onto what appeared to be a three-legged milking stool. Why would a milking stool be here, of all places? Liam held his breath, fearing it would tip over but Lizzie was experienced in the balancing act and all went smoothly.

Later, Liam with real appreciation realized that Lizzie was the self-appointed social leader, the doyenne of Bollard Lane. She exuded dignity and self-esteem and the neighbors knew it. Many came to her for advice and she was never without some. She even occasionally shared a cake of hotel soap and a purloined tea bag. She never explained her past to anyone so her presence was still a mystery. She wanted to keep it that way.

The teabag, though she preferred good loose tea, was passed from cup to cup. It had been used a good number of times before. The spoon stirred the sugar and, sure enough, the somewhat curdled cream of uncertain age that Liam managed to forego. The watery tea was consumed with something less than relish by her guests. Lizzie herself performed all of her hostess duties with great aplomb, her pinkie finger pointed skyward from her own cup, a tin one with a large dent in one side. She explained that she needed no food or extra clothing because her son, Michael, had a grand job and would be home soon to take care of her in fine style.

"I don't need a thing. My wee boy will be here any day." She had said this so often that now she believed it herself. She kept a sort of pallet of ancient blankets in one corner for his arrival. It seemed like a manger.

When Father Andrew and Liam left, they discreetly deposited several large sacks in a corner of the room, knowing that Lizzie knew they would and that she knew that they knew. Before they were all the way, huffing and puffing, down the steps, they could hear the rattle of the bags as Lizzie searched through them to find treasures like tea and chocolate and, praise be, a new tin of milk. Greatest of all would be maybe a pack of cigarettes.

From one of the sacks, she pulled a remarkable wide-brimmed red hat. Lizzie smiled, knowing that would surely further mark her as one of the aristocracy.

Father Andrew called Lizzie 'Mother Ireland', after the literary character Cathleen ni Hoilihan, the embodiment of Ireland because she showed all the attributes: she put up a good front to all, she was ready to fight for herself or for Ireland, whichever came first, and she never missed a chance to sing or laugh. And she never gave up hope.

Liam was never quite the same after that day. They visited many, some of whom were ill and he hated to inhale while in their digs. He saw hopelessness but he never forgot Mother Ireland. He never forgot the Catholics or the Protestants who needed more than a good bombing to pull them from the gutter.

CHAPTER 20

Mousy Rebel

The second son, Brandon, pretty much faded into the woodwork. He was largely ignored by Moraine, who had more urgent matters with the older children. He was passive and needy and spent most of his baby years in a wet nappie, with a runny nose and only one sock.

He had his mother's mousy coloring and later there was an excuse for a mustache that he had grown to make himself look older. His eyes were a watery blue. He bore little resemblance to his flamboyant father. He fully looked the clerk that he was. A fine occupation, surely, but somehow one that stamps its owners with the indoor complexion, the unglamorous mien that attracted so few young ladies. Brandon longed to be a flashier type that he admired because of their appeal to the ladies. This was a situation that endowed him with the need to find excitement at the racetrack.

"Good morning, sir!" With a touch at the peak of his checkered cap, the tout, known imaginatively as 'Horse', sidled up to Brandon.

Such music to Brandon's ears, this being called 'Sir'. At least, here at the track, he was accorded proper respect. His left forefinger self-consciously stroked his nearly non-existent mustache; his right hand tugged on the bottom of his flamboyant vest. The vest, a pinstriped number that seemed to be a shiny purple - shall we call it aubergine? - had been bought with part of the money owed Moraine for the rent.

How splendid and noble Brandon felt as he strode - not walked, mind you - to the window to place his bet. He never could discern whether he was the more accepted here among his jolly friends or downtown at the turf agents where he was so cordially welcomed. Accepted is the key word here. For a lonely young man too wimpish for his own mother's taste, never a magnet for young ladies or other youth for that matter, Brandon had tried to create another world for himself. He had set his sights rather too low.

With his winnings today, dependent on the fleet feet of Rock Beat, by Rock Coast, out of Dune Maiden, he would pay off his mother and purchase a new hat at least as grand as the vest. He would wear that hat cocked smartly to one side. Rock Beat had an impeccable background. With the new Panamanian jockey up, how could he, Brandon O'Shea, lose on this sparkling day? Not even the usual misty, drippy day.

Well, as it turned out, he could lose. Despite Brandon's hoarse whooping and hollering, Rocky plodded on. He was a mudder and didn't know what to do with a dry track. Rocky might be plodding on to this very day, but Brandon didn't stay to see. He literally shot out of the stadium (maybe he should have replaced Rocky in the race), trying to disappear before he had to pay off side bets with money he didn't have.

The one place he really wanted to be, the one place forbidden by this circumstance, was nestled up to his mother, being reassured that everything would be all right. Now, where would he belong? She probably would ignore the whole thing, anyhow. Not many hugs, there. Hugs from his Ma had always just been a fantasy. Besides, he was too grown-up for such fantasy.

He knew from previous experience that there would be no more hat tipping, no tugging at the forelock, no greetings from fellow bettors. Was a man no more than an x-ray? Just skeleton and a wallet showing and if the wallet showed empty, nothing.

Very little of his sparse wages had ever found its way into Moraine's reaching hand. He affected a smug look and an attitude of 'I've earned this' to combat her constant whining requests for his share of the rent. The share was not totally necessary and he knew it because his father had a job.

Moraine looked upon the payment as a discipline and, had she been able to get her hands on any of it, she would have saved it for Brandon for later. She repeatedly told him to join the police force so he would make more money and maybe get a chance to get a whack at some of the Protestants. It would be a steady income and he would wear a uniform, saving money on clothing, not to mention the saving of the world from the sight of aubergine vests.

Instead, he was drawn toward clandestine meetings with the Protestant segment of Belfast's conflict. He truly felt more comfortable with the status quo and had no gleaming desire for independence or joining the Republic. He wanted to be 'accepted' even though he was one of those 'mixed' offspring. His solid job with Her Majesty's Royal Postal Service might no longer be so solid if the picture changed.

He was, however, not too enchanted with being called a 'dirty Protestant' by those of his acquaintances who noticed him at all. He had begun to go to meetings with those 'dirty' ones in order to have some affiliation of his own. The horse races offered him only a limited amount of belonging since he had so little to wager- even less after the Rock Beat episode - and he almost always lost that.

His meetings with the Reverend Paisley radical groups brought him into the charismatic circle of the huge man's followers, all ready for conflict. The Why really did not concern him at first. The flaming speeches about the greatness of the Orange Order and the glorious victory at the Battle of Boyne did not touch him at all.

What drew Brandon and really gave him resolve was the notion that he could belong to such a group. He rather liked the image of himself in an orange be-fringed sash, with bowler hat cocked to one side, swinging his bumbershoot to the rhythm of the *lambeg*.

It would be sure to get him some attention; his Protestant Da might approve, and it would drive his Catholic mother bonkers. Well, hadn't she always pushed him to participate in sports like 'other guys'? Not to be different? Hadn't she tried to control his whole freaking life? Herself with all her talk about being peace-loving. She was a walking tempest in a teapot.

These were thoughts that had never before entered Brandon's mind. For the moment he understood about false fronts and how his own mother might be more warlike than anyone he knew, but the moment passed and his clarity fogged. It was a long time before he could see that he was doing no earthly good in the meetings and plottings. It was his social life. It was his time to enter a room where everybody knew his name and might clap him on the back, even buy him a Guinness.

CHAPTER 21

Delia

Brandon's first encounter with The Party was not heralded by trumpets. Feeling like an international spy in the films, Brandon looked both ways, tore around the corner and scuttled into the Orange Hall on George's Street. It was so barricaded with barbed wire that he had trouble finding the entrance. The doorman was not prepared to let him in because he had no ID and no one was to be trusted. Finally, a friend of Brandon's father, a member in good standing, persuaded the doorman to 'let the boy enter.' The doorman still kept a suspicious eye on him.

One thing went the way he had hoped. The men lolling around the smoky, beer-smelling lodge room gave him the once-over and grudgingly allowed as to how he could stay around, inasmuch as he was Patrick O'Shea's son - and Paddy being a darling man. Still, suspicious of his intentions, all kept an eye on him and queried him very closely. His mixed heritage was generally known and they could not be sure he was not a spy. Three men removed themselves from his company and retired to a corner to grumble about this newcomer. He was still a loner but he didn't feel that part.

It was in the company of one of his newfound pals, with whom he was having a wee sip, that he met Delia and was swept off his feet.

After a few meetings at a nearby pub, he decided to bring her home to meet his family.

Whatever possessed him? He expected open admiration and astonishment at this butterfly. He was, himself, a mite hesitant about his family being worthy of this elegant, sophisticated creature.

She openly adored Brandon. How he basked in her reflected glow! He had needed this attention all his life, but had not learned to discern the sincere from the phony.

Delia was quite tall, taller than he by a bit, even without her trademark spiky heels. Her hair was red, perhaps originally, but enhanced to a mahogany shade, while on her fashionably spare body she wore only the most modish and becoming clothing. 'Modish' may be translated to say 'short'.

There was a little something, though a quirk. Every time he was with her in her minuscule apartment and the phone would ring, she would jump nervously, rush to answer and turn an ashen color.

Delia was convinced that an ex-lover who could not get over her fatal charms bugged her calls. When she and Brandon walked down the street together, she was constantly looking over her shoulder for 'someone' who might have been shadowing her and she feared that she had put Brandon in mortal danger. This feeling was not alleviated at all when she noticed the lace curtain lifted and dropped at Mrs. McBride's house as they walked into his house. Mrs. McBride was 'on duty' and Delia was convinced that this was part of an international ring.

Delia did have a fine point and that was her genuine affection for people. Everyone she met interested her. One facet was her rather pathetic need to be liked herself that formed her subconscious link to Brandon. She was a good listener, and the tales she heard were often sparked by some trifle she shared about herself.

She was particularly proud of her roots. More often than necessary, she dropped a few words in conversation about her brilliant, highly decorated and dead father, plus all of his shining forbears for generations past. The tales were not necessarily true and they were invariably boring.

Her charm stemmed from her ability to relate small anecdotes such as breaking her heel (3 inch, no doubt) that sent her for help to the first doorway, which turned out to be a massage parlour. They offered her a job. It was easy for her to laugh at herself but it was unacceptable if others poked fun at her. That, eventually, was where Brandon blew it.

Overconfidence, occasioned by the newfound intimacy, gave him the bluster to begin to tease. Greeted by huge laughter from his family and sick smiles from Delia, he stepped over the line when he smirked, "I reckon the Four Star General would not approve of the family roots. Your roots are beginning to show brown. It's about time for a visit to your hairdresser, isn't it?"

A whirlwind passed by the nonplused Brandon and disappeared. He never even saw what hit him. Nor could he figure out where he had gone wrong. Ever. A little obtuse, that boy.

CHAPTER 22

Facing Difference

Dear Bridey:

   I don't believe I want to hear another freaking word about Boston right now. You have absolutely no view of how uptight I am. When I tried to get a cab yesterday in the rain, they all just kept passing me by. Bunch of Protestants rule the world out there. I was soaking wet when I finally crept onto a bus heading for the Center City. And me dressed in me best duds. With puddle water splashing all up my legs, cold as ice. There was a truck unloading huge cartons on the next corner and when one of them dropped, making a noise like a shot, I went up in the air about six feet. It's that panicky I am, since yesterday's car bombing and the death of a child.

   When I finally got a cab, there was a radio program on that kept calling the Pope the 'Antichrist' If that were not bad enough, I could feel the cabby's eyes in the rear vision mirror, watching for my reactions. I tried to keep a poker face by staring at the back of his head where that stupid cap tilted on his head. Then I tried putting my mind on the cracks in the plastic upholstery. Then I put my entire attention on a spotted dog that was trotting down the street. Finally, I stared at a fat lady with an orange raincoat and matching hat. But I could feel my stomach churning and by the time I got to the dentist, I had to upchuck in the loo. I just know I will have my nightmare again. Thanks for listening. Mor.

Dear Moraine:

Do you think that if you were more aware of the peaceful activities around you and if you got out of the pack-lunch-walk-to-school routine once in a while, that your jitters would settle? It has begun to dawn on me that I can do things for myself so that I won't be so frustrated and angry all the time. Where is that anger to go?

Today, my youngest son said to me, "Ma, don't you like anybody? You act like you are afraid for me to go over to Jose's house to play. Why don't you come with me to meet his mother?"

From the mouths of babes comes wisdom, so I did go, ready to disapprove. I found nothing to disapprove. We are much the same, especially protective of our children. Elena and I giggled and swapped recipes and complained about our husbands. It did a world of good and I believe I have found a new friend. That's why I thought it might make you feel good to join in some activity with other women in the neighborhood.

Love, Bridey

CHAPTER 23

Meeting Ground

Moraine had been trying to get to the center of Belfast, to the Europa Hotel, which was the terminal for all of the buses. In order to get to Armagh, where she had heard there would be a meeting of just women discussing their health care and having tea together, she had to take a bus. Rebels on the highway were stopping even buses.

She had left the dirty dishes and unmade beds to be sure to arrive on time. There were similar meetings going on right in Belfast. One was held on Tuesdays not too far from Springfield Road. The problem was that such meetings frightened the men. They just could not be too sure what women would do at a meeting of their own.

Even in Armagh, vans were sent by the women's organizations to safely convey women to the meeting place. The meeting places were very low key and often had to change locale. The women knew where they were but all feared an incident with one of the militant groups or a disgruntled husband. They were between two forces. Slowly but surely, the women's groups were becoming ecumenical as well. This increased the danger level from the outside by 'religious' protesters but gave a warmer feeling of support from within. Moraine wondered why should there be so much intrigue for such innocuous reasons.

The attendees were all amazed by how much they all had in common. There was no *danse macabre* as each tested the other for all directions of attitudes. There was, however, a trust-testing period where each tried not to stare at the other, or at least not to appear so.

Her uneasiness lasted until she was seated with her first cup of tea in her hand. Although she had come alone, she found herself sitting with other women who had come alone too but who were cheerful and friendly. They began by looking at each other's dresses, then hairdos, then finally into each face to find a friend there. All were there for something and all got more than they expected.

"What did you do with your children today, Bernie?"

"My mother is taking care of them just so long as I get home by 3 o'clock. My father would have a fit if he knew why she is keeping them for me. He thinks I should be home waiting for Kevin to come home for supper and that he would have a fit as well. If the bus doesn't run, she and I are in deep trouble."

"I brought my baby to the day care. It was good last time even though I was worried about leaving him in the care of strangers. It's hard to learn to trust, isn't it?"

"I had to come by Dublin Road and I heard that a car bomb was discovered this morning in an old car beside the road. We were detoured and I thought I was going to be late."

Inevitably the chatting turned to the latest violence which involved a child burned to death in a house put afire by black masked persons who objected to a mixed marriage. Moraine felt sick but luckily the meeting started and she was diverted.

The leader came to the front of the room, which was an old Sunday School room in a Presbyterian Church. It was announced that at alternate times they would have to change the meeting place to keep any outsiders from guessing their location. At the moment, the next site had not yet been selected. Also, by having different venues, there would be opportunity for more women to attend. There was a lot of talk about how to spread the news about where the meeting would be. It was decided that a list would be made and a phoning chain begun. Even if a woman had to go to a pay phone, she was bound to call the next on the list until the message was passed.

The speaker began, "This morning, we want to talk to you about your nutrition. We need to have an idea of how many carbohydrates you and your family are consuming because it may be too much. It must be kept in balance with your proteins. Sounds boring, doesn't it? Wait until you hear Trish."

She was right. These women who had taken little notice of their own health and had expected to grow old just as their mothers, were transfixed by all of this information. Calcium was a big item because most of their mothers were bent over from osteoporosis. The care of teeth to avoid false teeth, which they considered inevitable, was included in the advice.

They had been busy exchanging views but also they had been busy crafting wreaths as they talked. When they took a break, they admired the herb and dried flower wreaths that they were making for their front doors. Moraine wondered how she would tell Patrick where she got it. It was right much lopsided, a little heavy on the thyme branches, but she was proud of it because she had rarely tried to make anything with her hands.

She noticed that a few of the other women confided how angry they were about their home life and the fact that they all had to appear cheerful for the children. She had been doing that very thing. Now she wondered if it was a good idea to always be presenting a false front. Children always catch on. That was one way the violence seemed to simmer on under the skins. On the other hand, it would be better to find outlets for those frustrations. The organizers of the meeting had known that the handwork on the wreaths would not only be a way to get the conversations rolling but would also demonstrate a good outlet for frustrations.

Plans were made for the next get-together but best of all, some of the women planned to have tea at each other's houses before that time. Friendships were building among the women who did not go out to work. There was a little relaxing of the eternal tension. Would it be possible that, in time, women would really become peacemakers even on the deep insides of their feelings?

CHAPTER 24

Bonaventure and Ruby

"So, when the High King decided to take himself a second wife, his first wife, Edni the Proud, was infuriated. She decided that instead of physically harming the new one, Cairde, she would simply get her out of the king's sight. She sent her to a mill where Cairde was forced to grind flour by hand, often too tired at night to entertain the king. Edni thought she had found a way to get rid of her rival."

It is strange how a terrible episode could change things for the better. Moraine had become obsessed with pure fury and was looking for revenge. One of her 'friends' told her that Patrick had been spending extra time at the pub and had been seen walking the redheaded barmaid home, laughing and practically prancing. Of all things, he had stopped even trying to hit her even when he was drunk. Could that possibly be a bad sign?

When she confronted him, he just said, matter-of-factly, with no attempt at deception, that he had been lonely.

Came her retort, "And just what do you think I am entirely?"

"It really wasn't anything, Maureen, *mavourneen*. You are my wife and the only one I'd be loving."

She wasn't satisfied. It was just further proof to her that she did not measure up. Measure up to what? - her own expectations and her own nar-

143

row life? Some more of the glacier slid away.

She began to read and think about marriage and decided that it had not been designed as a trap but as a partnership. For her, that was news. She felt it was a real breakthrough, just for her.

Her obsession with this 'failure' stayed with her, though. She was obsessed with trying to improve her looks, her health. She overlooked the fact that what she needed to be was more just plain fun and understanding.

The intensity of her perceived feeling about the barmaid simply grew. Prayer did not help as much as she had hoped. For one thing, she found herself praying for something to happen to that 'redheaded tart'. Then she was afraid that something would happen and that it would be her fault.

She tried to change her mindset. She vowed that there would be no more obsessing. The fact that her prayers resulted in changes in her own perspective didn't occur to her, but she was even nicer to the cats.

After some months, she did find that a large slice of the problem had been her expectations. She needed to form the kind of partnership that would make her own life satisfying and less critical of Patrick.

Father Reilly was not the one to be thanked for all this new thought. At the library, she had made a friend, a nun named Bonaventure (known as Bonny) who understood. She continued writing letters to her cousin in Boston. She found her own therapy.

There was more to Patrick's apparent need of a confidante than just flirtation. Unknown to Moraine, his job at the factory had dissolved.

Through various pressures, political and otherwise, the American auto manufacturer went into bankruptcy, thus destroying any satellite Irish firms that had built on the fringe of the expected prosperity. The auto parts business was shut down and the previously secure workers were on their own, jobless. They were promised training for other jobs, but there were no other jobs.

Patrick was mired in a deep depression and hopelessness. His job had given him great self-esteem amongst his cronies at the snug. They could sit around a pint, make sport of the boss, and tell, with frequent nudges and winks, funny stories purported to be about him. It was as though his props had been knocked out. He called on every company that was advertising for help in the papers and a couple he just heard about but to no avail.

Nothing he had relished before had any savor now. His mouth was dry as dust and his favorite ale did not quench. One of his best mates shot himself. The suicide was a wake-up call for Patrick. There was a wife and children left helpless and he did not propose to do that to his family.

It began to appear that he would have to take a job in London, probably in construction, and for practically nothing by way of pay. The English would look down on him. He could only come home a few weekends now and then. Of all the swirling thoughts he had, he mostly dreaded telling Moraine. His own self-image had always shown him as the strong wage earner on whom Moraine depended for her very life.

He could not tell Moraine. He knew that no paycheck would be forthcoming. He could not face the railing he would get. He never even suspected that she could be sympathetic, that she would understand. To him, it would be like telling his mother that he had fallen short. It was much easier to talk with Ruby who was working in the pub and knew what was happening. Besides, Ruby had no expectations of him.

Every week, Moraine went to the library where a whole dozen tots waited for her to tell them stories. Her storytelling was well known all about the county. She was much in demand at gatherings and fairs.

"When Esirt the poet went back to his king, Iubdan, he told him of the promise that his king Iubdan of the lepra would go to meet the king Fergus Mac Leide of Ulster. Now the little people are not much in favor of mirrors, not even looking into still pools after the rainstorm. For this reason, they do not realize how small they really are. In their minds, they are the same size as everyone they see around them.

So, the first thing that happened when Iubdan and his wife Bebo went to visit the other king was that they had to wait in the great hall until he came. The little man peeped into a bowl on a table and fell in. He fell into a bowl of oatmeal that he said later tasted good. But when he tried to stand up after slipping and sliding, he found his little feet were stuck and his good wife couldn't do a thing but search for help."

"And now, children, you will have to wait until next week to find how he was rescued."

"Aw, Ms. O'Shea, tell us what happened. Please, please, please."

But Moraine was all told out and she went out of the children's room to find a hiding place.

At the library, when she was hiding in the stacks to do her weeping, she was surprised to feel a kind hand on her shoulder. When she straightened up, wiped her eyes, and started to scuttle away, her eyes met those of a nun, just about her age. Of course, nuns always seem to look younger than their age. And her eyes were kind eyes.

"You seem a bit down, dear. Can I help? I am Sister Bonaventure and I would wager that you have man trouble."

"How would you know that? It could just as well be the state of this bloody country." In some anger, Moraine pulled away.

"We all have some relationship or other that we are struggling to maintain. In the convent, we are trying to love our sisters, our Lord, the rules of obedience that we can hardly abide. Sometimes we even become infatuated with a man. That, we must learn to work out."

"A man! How could you?"

"Have you never seen the beauty of some of our priests? And in some instances, the beauty of our nuns? Did you think our vows would relieve us of temptations while we are young? It is because of our humanity that we are able to help when we see others in distress. Don't deny us our humanity. What is your name, dear?"

In spite of her efforts at self-control, Moraine began to snuffle out the details of her obsession. As she talked, she came to realize that she was the only one who could change herself. She could not change the situation. She even confided to Bonnie the reasons for her name change from Maureen to Moraine. A good listener, that Bonnie.

"Do you think you are the only human with this problem?" said Sister Bonny. "First, you may need to learn a little short prayer that may help. It can be used as a mantra. 'Lord, I know that although I am unique, my problems are not.' Somehow, that gives one the right to rejoin the human race at a time when one feels so separate and low."

At home in the welcoming arms of the old green chair, nose and eyes still red and puffy, she sat thinking of her day while the silence grew. Peace flowed through her for the first time in days. She was drained. Her heart and mind slowed. She only wanted to bask in this glow of warmth and caring. Maybe when she started thinking again she would have some answers to her quandaries. She felt as though she were at prayer. She was.

She stroked the silky fur of both her cats in equal measure so there would be no jealousy. Bel and Sam were named for pagan festivals, which used to be held before Christianity. She didn't exactly admit this to Patrick or to the children, or especially to Bonnie, but that part of her heritage, the days of Queen Maeve, kept popping up. She had a wild desire to build a bonfire and to dance but she kept to her new quiet.

Sister Bonnie and Moraine became such regulars in the Reading Room that the librarians became accustomed to the little murmurings in their corner - and also had a little gossip about their relationship. Lesbianism was, to them, a dreadful and mysterious affliction. The fact that Bonnie and Moraine were able to communicate on a feminine level never occurred to them. It was more fun to talk about them and snicker. After all, women did not have such close friendships that these librarians knew about. But Sister Bonnie became Moraine's *Anam Cara*, her soul-friend.

When they talked together, Moraine began to see that she must become a separate person, one who had her own present and future. Most of all, she needed to be less serious and intense.

She needed to be more like fun-loving Paddy, of all people. Most of all, she needed to give more mindful attention to herself and to her family and to trust her intuition. She had a strong intuition, a knowing, that she had inherited from all the generations of women before her. Strangely, one of the Irish words for the 'knowing' is the 'glamour'. She had no feeling of glamour. Further, it seemed that she needed to find ways to rid herself of her fury.

Bonnie thought prayer could open Moraine's inner angers and relieve the pressure. Maybe so, but a good brisk walk sometimes worked even better. Moraine needed to learn that being 'nice' on the outside while seething on the inside, could do as much harm as good.

Bonnie told her to lighten up, to laugh more, and to dance, even if Patrick wouldn't dance with her. Maybe in time, he would. Dance in the kitchen and in the garden while watering the flowers with the grandchildren. Bonnie thought the Women's Meetings were a great idea and she wanted to go with Moraine the next time.

Moraine's plot to cause the redheaded one to lose her job began to seem childish to her. She no longer waited on the corner where the couple might likely walk by, although they never had. It seemed to Moraine that she had been giving her anger too much space and energy

   'Not show anger? OK, Virgin Mother! What do you make of that?' She danced and swirled in the kitchen and tangoed down the narrow corridor using a pillowcase for a mantilla.

CHAPTER 25

Night Out

Moraine had made a decision to talk Pat into going out together on a date, maybe once a month. True, it had been a long time since she even thought of such activity. "Keep Romance Alive!" she read on the front of one of the women's magazines. She hoped they could afford it and that it would give them some small taste of their old romance, some small spark of electricity.

Pat agreed, not at all reluctantly, because he saw this as an opportunity to show off to Moraine how well he was liked and recognized at the Cork and Clover where everybody knew his name. It would be a great lift to his ego and to his standing with her. He oftimes felt that she underrated his importance. She couldn't understand the importance of the Pub and his Orange activities even though she supported him in all his parades and shenanigans.

The date was made. Moraine was all gussied up and eager to go out for the evening. She had harped considerably about never going anyplace anytime, until Pat, in his dense way had gotten the hint and invited her out for a drink. Both of them felt younger by at least ten years as they started out after dark for the pub.

Moraine had dressed in her best and would much rather have gone someplace a bit romantic with candles and all, but Pat's inspiration didn't stretch that far. In fact, even though she had on

151

her best black high heels and her best long-sleeved black dress, she had to put a jumper on under her coat because *Himself* had decided they should walk, freezing though it was. The jumper was a happy circumstance because it was a cranberry color that gave her a bit of a blush. Black was never that becoming, but every woman needs a black dress for such as funeral going.

Her best shoes were destroyed entirely by her sinking fully into a puddle left from the afternoon downpour. By the time the festive couple arrived at the pub, Moraine was plenty warm enough for the two of them. Fact be known, she was fully irked and proceeded to down one glass of ale almost before it was set before her. '*Just to warm meself*', she told herself.

On the other hand, their entrance had been all that Patrick could have dreamed. The moment they were in the door, there was a great welcoming from all present. "Ah, there's Patrick, me boy." "Come over and sit with us, man." "We've been wondering where ye've been." Pat, Paddy, Patrick rang out and his swagger became visibly greater. Everybody knew his name! The barmaids all flurried about serving them and giving covert glances at his wife. Wives were seldom seen within these walls.

"We'll sit where Chrissie can serve us." Chrissie already knew what Pat would have and had waited curiously to find out if the wife would have the same. She would.

A further becoming flush from one more ale resulted, along with a noticeable loosening of the tongue. A splendid self-confidence ensued, giving Moraine the certain knowledge (along with the rest of the celebrants) that she was completely right on all fronts.

Pat made the opening remark, a perfect, if unplanned, icebreaker. "Well, Stormont's at it again."

Then with a cough and a puff, caused by years of smoking, Pat lowered himself into his accustomed seat. "Sit here, my dear." He expansively indicated the stool where she should sit. That did not set well with Moraine who wanted to be able to see all of the other customers who were now behind her back. She had also wanted to sit in a comfortable chair. Most especially, she did not want to be told what to do at all.

From a corner that was so smoke-filled that even the faces were obscured, came a voice that she thought she knew.

"If they'd get rid of that IRA bunch with their guns and explosives, we could have a little peace in July."

"And what about the marchers? Do they have to march right through our streets? Are they without blame?"

"What makes you so sure that is the problem? Any *eejit* can see that if they give up their arms, Paisley's people would walk all over them. Not a single job for a Catholic. Not a decent wage with all of the factory closings."

'*Ooops! What was that about closings? Close call*', thought Pat.

"So, you think with that magnificent thinker of yours, that there'll be plenty of work if the U.K. drops us. You believe that the losing of the post office and telephones, all the civil jobs won't make any difference?"

"We'd be part of the Republic at last without all of this division."

153

"They won't have us. Dublin looks down on us and don't want their apple cart upset by a flood of poor Catholics pouring into their census."

"More women elected to Stormont'll do nobody any good. You never know which way they'll turn, what with hot flashes and all."

Moraine was touched to the quick, infuriated. Emboldened by her ales, she came to the rescue. She had been busily concocting clever answers during the whole dialogue. "They're doing a splendid job altogether. They know how to negotiate, which is more than I can say for that lazy lot o'men. We even have to have an American come over to work for us to make peace."

Pat's lovely glow dimmed. Here was his wife, sticking her nose into men's business. How could he get her out of there without letting others know what a mouthy one he was connected with?

The trip home was worse than the beginning voyage. At home, each stomped to a separate corner and glared until they had to give up and go to bed, clinging to opposite sides of the bed. Before snoring relieved the impasse, Pat had considerable to say about her infraction of the unwritten rule of the pub - women are to remain decorative with their mouths shut. After all, what did she know about anything? She asked herself the same thing and resolved that she would know as much as possible about everything.

Moraine thought she heard an audible splash as a monster berg parted from her glacier. She wondered if the noise would awaken Patrick. Then, in her dreams, she watched the ice floe as it floated away.

CHAPTER 26

Grannies

Dearest Bridget,

How in the world are you? I think of you in Boston, all safe and sound in America. It must be lovely to be able to go out in the street and not have, in the back of your mind, the fear that there might be a shot or a bomb and to be able to send the children walking safely to school even though you still go with them.

Only Thursday, when I was walking with the children, one of the soldiers had the nerve to speak to us (we pass the same ones daily), so we threw back our heads and marched on. Now that I have thought about it a little, that soldier is little more than a child himself and must be lonely posted in Belfast where there is so much hate. Perhaps I'll take him a bit of cake next time I'm going by. Just my luck, Mrs. McBride (remember her?) might see me and report to everyone.

There has been no problem lately but you know there is this little pulsating nerve somewhere in my brain, anticipating a crisis, especially being married to a Protestant. You know, Love, I really believe that we Northern Irish are so accustomed to that feeling that we would have a great void if it were gone. How do you deal with that relaxed feeling that living in America must give? I really believe that you don't have the 'crisis syndrome' working for you.

You have not heard much about our daughter, Angela, that I didn't much want at first. Of course, she is no longer exactly a wee one. She is twelve now.

See how long it has taken us to really get back in touch! Even though we have written to each other, I haven't had time to really communicate because I have been awash with children and grandchildren, disputes and all the problems in school.

I really wanted to name her an un-Irish name. I picked Tatiana. Doesn't that sound exactly like a queen or a princess or a glamourous international spy? Although I try to be fair with the other children, I find myself buying sparkly things to go in her hair and ruffly dresses and even socks with ruffles.

Patrick wouldn't put up with that Tatiana stuff. She is an angel so we named her Angela. She has the most awful asthma. Sometimes we have to take her to the hospital and it is so scary that I stay with her.

Angela is about the same age as Kathleen's two, Maeve and Kevin, who live with us. Rather a full house, just when Paddy and I were beginning to speak with each other occasionally. Probably it was the 'speaking' that did it - another child and a sick one at that.

I never hear from Kathleen. The whole situation just makes me so angry I want to hit something. She seems to be quite content without her children clinging to her.

Come to think of it, you have grands too, but they don't live with you, do they?

They tell me that spanking has gone out of style but, believe me, I do a bit here and there on the deserving behinds of our three.

The barricades are still up on Springfield

156

Street. Do you remember when we took buckets of paint and put huge brush strokes on the store walls? Murals they were not, but they spoke of freedom. I thought that peace would have arrived by now like flights of white doves. You and I were ready to fight for it! Perhaps fighting was the wrong direction. Could we have made things better?

Father Liam is turning out to be a better man (and son) than I had anticipated. He has learned something about peace that I never could have taught him, maybe by default. He really believes that Northern Ireland will have peace if we all work together! Hah! He now wears a threadbare cassock in spite of all I can say and he tends to the people in his parish as though they were truly his children. Where did he learn that? After all of those stories from the Old Testament that I told him nightly, I would have expected him to be more rigid, more willing to go out in conflict with the mixture of people he meets. He says that he learned, not directly from his abbot, but from the first time he saw real need. He saw people who have such great physical needs that they can't bother to think about politics or conflict. They can hardly pray when they are hungry. He is quite an influence on our Angela who may enter a convent as soon as she is 16. Sometimes I think she is looking for an escape instead of having a real calling. But, I thought Liam was escaping and now I wonder if he did not have a 'call' that was subconscious.

So now, Kathleen is away with her man working with the IRA, secretly, of course. I just don't feel sure she has considered what she is working for and I can't imagine what the work consists of. Are they moving firearms and explosives? It is frightening. She is in danger of being recognized and put in prison and the whole family would have to suffer. If I only felt that she knows why she is taking these risks, it would be more bearable. I hope she knows how to fight.

The streets here are still patrolled, especially the Catholic sections. I am very cautious when I walk my grandchildren and my Angela to school. Our estate is right on the border between the Catholic section and the Protestant one. We are not acceptable in either. We are the mongrels of Belfast. In fact, I suppose we live on the border in many ways, philosophically as well. I always walk with them but what good does that do except to make them more afraid? If anyone dares to harm them, I will get them. Kill? I will if necessary. I have told them that nothing will happen to them! They know that I can really prevent nothing.

I have whined enough. We are healthy - and did I mention that Brandon finally won a small bet at the races? Maybe that's a good omen.

I never see him any more. He is out every night and I have no idea where he spends his time. Very frustrating. When I taught my children to be independent, I didn't mean for them to leave me out entirely.

Please write soon to me and tell me more about America.

Love to all,
Maureen (Moraine)

1083 Kelly Street,
Apt. 431
Boston, MA, 43021,USA
October 11, 1990

Dearest Maureen - you are Moraine all of the time now?

What in the world makes you think that we are living in a haven of peace? The streets where we live have no visible barricades unless one counts the yellow tape that surrounds a house where there has been a shooting or a stabbing, and that is too often.

Evictions are the worst part. People who have no resources are evicted from their homes just so a fancy new building can be built on their home site. Of course, in some ways, the site is like a ghetto.

Most of our near neighbors are Irish or at least of Irish descent and we seem to have transported the same bigotries toward each other that we would have had on the Auld Sod and the same ways of dealing with them. There are blacks and Italians, Asians, Hispanics and who knows what all. At least in Ireland, we were all Irish without all of these other cultures to deal with and compete for jobs with.

Some days I wish I had married an American - or not married at all. I think I thought I must be married.

Women aren't like that any more, are they? Irish women have learned to be independent. It must have taken an enormous effort to learn not to be like their mothers. Many come over on their own, earn their own living, and don't have babies every whipstitch. In fact, that has been true for decades and I feel as though I missed the boat (no pun intended). There is always somebody who doesn't get the message and this time I was the one.

David has become accustomed to being out with the 'boys', so that I do all of the discipline of our two grandchildren. Did I mention that our daughter, Doris, is divorced, works two jobs and cannot afford day care? She can afford me, however. I love the children, but they make me very tired. I would like to smack them to make them settle down but there is so much watchfulness in the schools and elsewhere about abuse that I might be blamed for much worse.

Like you, I walk to school with the children. The buses frighten me with all the older tough children and even though I have tried to teach them to fight back (including some dirty punches), I'm just not sure. The streets we walk through have a lot of bums and drug users hanging around. I have started carrying a baseball bat under my coat and I wouldn't mind having a set of brass knuckles (illegal here).

The worst part of this full-time grand mothering is my lack of energy. When I was younger and raising my own children, I had lots more energy and patience (still not enough, though). Now with these wee ones, I have a tendency to give in, to spoil them, and to let them go on with their shenanigans. This is not true love. I am just too blooming tired to be consistent. Shouting about turning down the volume on that horrendous noise they call music doesn't do a thing about bringing the generations closer together.

The only thing I have learned over the years is that I don't have to be liked by them or anybody else.

I do love them so much. They give rise to all of my protective instincts.

Because I take care of the children, I don't work outside of our home. I miss the old community feeling even after all these years. My neighbors are almost all gone to work all day. In an effort to avoid just housekeeping and waiting, which just drives me crazy, I have decided to start back to school at night because that way my daughter can take care of her children. She is not too anxious but I have told her that's how she will pay me for my baby-sitting services. I've always wanted to learn more about everything. Perhaps I will take courses such as Peace Studies or Conflict Resolution. Sounds ludicrous, doesn't it, when I really want to pound on someone or something?

I guess it's because you are clear across the sea that I feel like I can tell you things from inside me - how I'd love to go dancing or how frustrated I am in our marriage or how tired I am of being uneasy. That may be the whining you mentioned in your letter. You may turn out to be my therapist.

Does Lily Cassidy still live in the block? Are her roots still dark two weeks out of the month when she hasn't had her blonde dye job? They must be gray roots by now.

Affectionately,
Bridget

CHAPTER 27

Lurking Banshee

Maeve and Angela now shared a room, to the disgust of both. Kevin, the lord-it-all male, had his own empire.

Moraine tried to remember the last time they had tolerated her story-telling time, her time to tell them the glorious stories of the Old Testament: Joseph and his brothers who tried to get rid of him, Delilah and Samson, the man she overcame, John the Baptist and his head on a platter. She loved all of the slaying and smiting. Better still, the stories of Finn McCool and Cuchulain. Best of all, she loved the warrior woman who taught Cuchulain to fight.

At least they were in the house and asleep. They were inclined to stay out at night past her curfew. After all of the years of walking with them every time they went to school or watching them as they played in the fields, Moraine was still uneasy when they were out of her sight and control. With each birthday, they grew more difficult to control.

All of the mothers and grandmothers had hovered over the children who now thought it silly now that they were totally grown-up and knew-it-all.

Moraine's uneasiness had been transferred to them although they always acted cocksure.

There was still too much conflict in the streets for them to be safe. There had been the reprisal murder of a taxi driver just last night and so it went tit-for-tat. When two children were burned to death in a fire set because of a mixed relationship, Moraine could not sleep for a week. First a Protestant and then a Catholic, eye for an eye.

The prison, Long Kesh, is a breeding place for violent reactions, with all the prisoners from both sides thrown together. Now, there are more than two sides, what with splinter groups forming from the basic ones.

There were only men imprisoned there but the roots of their violence are deep, from their family histories, from the politics of the land; learned as much from their mothers and grandmothers as from the situation at hand.

In the women's prisons, there were the same disagreements, stemming from anger and fear, sometimes the residue of an old loyalty to tradition. Oddly enough, in the women's prisons, there was occasionally an appearance of peace and understanding. Women seemed to carry within them some knowledge that the current situation wasn't working. In some cases, this was only a surface peace and the underlying feelings still festered. One of Moraine's old school friends, just released from prison, told her that she only became peace loving when she realized how very, very tired of fighting she had become.

Moraine slept fitfully and dreamed disturbing dreams. She dreamed of a huge black carriage with black horses practically flying through the air. This was not her shattered window dream. This was a brand-new dream that seemed totally unrelated to any current happenings. There were sound effects. The wild shrieking she heard chilled her spine and made her heart beat wildly. Could that be the sound of the banshee her grandmother had so dreaded, the harbinger of impending death?

Moraine awakened in a state. Her blankets were all off and she was freezing. The eerie sounds echoed in her mind. Quivering, she stumbled to the kitchen to start the tea. She could not shake the overwhelming sense of doom and gloom that stayed all day despite the sunny weather. She longed to see Angela.

The moment she was sixteen, Angela had become a novice in the Convent of St. Teresa.

CHAPTER 28

St. Catherine's Wheel

Angela had been a merry wee girl and her busy mind was often full of fantasies. Perhaps because it was the only exotic image presented to her, she had a travel-brochure vision of the interior of the convent of St. Teresa. Her glowing inner mind produced smiling nuns pulling from the ovens steaming loaves of golden brown crusty bread, beautifully aligned weeded rows of vegetables and flowers in the garden, sisters walking through the cloister side by side holding candles. In some ways, she repeated the visions that had lured Liam in his innocent beginnings.

The Sunday that she ran into the Mother Superior at Mass, she knew in a flash that the convent was the place for her. All of her dreams surfaced and in a kind of daze she dared to speak.

"Could I? -- Would you? -- Is there room?" Halting, she finally shut her mouth and gave the nun time to speak.

"Do you mean you would like to join us at St. Teresa's? You must realize, my dear, that you will have much work, more than you are used to. We really shouldn't risk taking one so young but we need new novices and I know you have been reared knowing your catechism. Come in next Tuesday and we will start you off."

The nun offered a quick prayer for forgiveness for her selfishness. It would be lovely to have this young girl to show off during the visit from the bishop and all of his entourage. She could bypass some of the formalities and get her in quickly. Even if the child didn't work out, she'd at least be there for a while, especially for the bishop's visit. There had been so few volunteers that had come to the convent in recent years.

When Angela raised the huge, well-polished knocker on the carved wooden double door, a tiny window slid open and blue eyes peered out.

"I've talked with Mother Superior and she said I could come now and begin my tests."

"Oh, are you Angela? Come in, my dear."

Once in, the picture began to change, fraction by fraction. The entry was about as dark and gloomy as it could be. Dark, dark aged wood formed every surface, wood that had been subjected to endless coats of furniture polish and never refinished.

To the left of the door was an ancient Catherine wheel that still creakingly revolved, sending the mending and food into the three remaining cloistered nuns. This really chilled Angela's spine as she thought about St. Catherine who was tortured on a wheel. Operated like a huge lazy susan when turned with a parcel on it, the creaky tray would turn out of sight into the mysterious caverns beyond with its contents to be rescued by the cloistered nuns within. They were very old ladies by now who had been unwilling to give up their cloistered life style when modernization set in.

No longer did the wealthy women come to leave their mending and embroidery on the old wheel. It was extremely rare in the midst of a city to have a convent either cloistered or not but these three would be cared for lovingly for the rest of their lives.

The sister showed her to the Mother Superior's office. Mother Anne was seated at her desk, appearing to be in complete control of an avalanche of paper work. Her serenity completely filled the room. Although Angela had met her once before, she was still shocked by the perfect beauty of this 80 year old woman whose blue eyes mirrored the sky blue of her in-house habit. There was kindness and love here, but a certain sternness of visage gave the sure knowledge that here was someone who would not be toyed with.

Mother Anne had papers ready for Angela to sign and some for her to take home to her mother and father for signatures. She was a little hesitant about the young girl's vocation and already had begun to regret her impulsive decision, but she decided to give her the tests. The tests consisted of hard physical labor, long hours of prayer and total obedience to the order. Obedience was usually the hardest for young girls to learn. Angela would be no exception.

The first assignment given Angela was hoeing the potato rows. She had not even known that potatoes needed hoeing, just buttering and eating.

With the straw hat and gloves provided her, she started out with great good will. Unfortunately, her stabs with the hoe dug up an entire hill of potatoes and Sister Ursula, who had been assigned to instruct her, lost patience almost immediately. The nun's expression did not change. The sweet face was still there, but the lips tightened and her neck tendons showed a little more prominently.

The potatoes replanted, Angela was reassigned to the kitchen. Huge mounds of risen dough faced her and she was directed to the table where the kneading went on. She stood between two novices who were showing signs of fatigue. Angela, once again, went at it with great good will, envisioning the yeasty smelling bread that would result.

In about five minutes or less, her shoulders, neck, arms were screaming for mercy. She had not learned not to go with all her strength, but to pace herself, to use just what muscles were needed. Learning that technique was no easy job. She was not relieved of this chore because there was no obvious mistake, but some poorly kneaded dough went to the bakers.

It was quite unusual for a young girl to begin with the labor part of her schooling. Angela had arrived on the scene when the convent was agog with expectation of the arrival of an entourage of bishops and a cardinal for the purpose of overseeing these women and assuring their conformity and obedience in all they did.

Her evening devotions on her knees in the cold chapel gave her a pleasant sense of martyrdom, which would have been even more satisfying had she been able to keep her mind on God Almighty instead of her aching knees. She would have understood Father Reilly completely with his arthritis. Her thoughts meandered around to concerns like her hairdo, her starvation (she had only a very light supper and no chance at snacks like at home), and her curiosity about the bishop's visit. Very little went into prayer. Later, she had an overwhelming sense of guilt at her lack of discipline. She feared she was the only one who had ever lacked self-control.

Angela did not even see her sleeping quarters until almost 'lights out' when she received a coarse sheet, pillowcase, and a scratchy blanket. She was then shown to a dormitory for eight (including her) novices. There was a water closet at the end of the room with a sumptuous supply of the coldest water she had ever felt. Straight from the bowels of the earth it sprang though she would have expected it to have a degree of the warmth from hell, being so close and all.

She learned to double her one sheet and she learned it just in time because the lights went out until four the next morning, the time for matins.

As the long days filed on, the obedience part was what riled Angela the most. Even when the nuns held their own services, a priest of the masculine persuasion had to be present for the sacraments. A male had to present the Host and the benediction. When she looked at the dignity and devotion of Mother Anne, who willingly humbled herself to these rituals, Angela was nonplused. Surely, this was not one of God's laws, this obedience to the priests. She did not understand.

This was the 'obedience' part she wouldn't get.

Before making her decision to enter the convent, Angela had given no great demonstration of piety. Perhaps she remembered hearing of Liam's first years as a priest.

She had just gone that day down the streets of Belfast and through the huge iron gates, knocked with the brass knocker on the forbidding oaken doors and was admitted for questioning by the Mother Superior.

Although the order was somewhat adverse to such inexperienced youth, the times in the country were chaotic (not to mention the impending visit from the bishop).

Angela had such a solid background in her catechism and rituals that she was accepted into the novitiate. Many of the nuns had known her in parochial school and knew her to be a gentle, malleable, and intelligent young woman.

They had never known her other side - that of anger and frustration because of what she saw as the hopelessness of her life in Ireland. What could she become? A schoolteacher? She was not trained for any useful job and there were not many jobs for her.

She returned home only a few times in the intervening weeks and Moraine was shocked to hear some disillusionment from her already. It was not just glamourous habits with their white starched collars and coifs that she had to starch and maintain herself; it was obedience. Angela had already discovered that there were teachings with which she flatly disagreed. What is all this pagan communion stuff? How come only the women were the servants? However, Angela was a determined, stubborn young lady and she went back dutifully without spending the night at home. That old bed had looked awfully inviting, though.

It was not so long before she smuggled a note to Brandon at his place of work. It started as harmless fun.

The outwitting of a 'superior' is always satisfying to those with no power. Of course, Brandon replied at once and, as the pace of notes quickened, she agreed to meet him at one of his underground meetings.

Even though her own brother vouched for her, there were those in the gathering who wondered where this young woman had come from. Of course she did not arrive at the meetings in her habit. She wore jeans and an old sweatshirt.

Some of those present were aware of their mixed parentage and even began to eye Brandon with suspicion. Their suspicion continued undiminished even after several meetings, despite Angela's apparent enthusiasm for the Cause. Trust was not the long suit in the organization.

"Ma," she had said, "Nuns are just servants in that house. Not just servants to our Lord and His Mother, which is why we are there, but to the priests as well. Did you know that no matter how hard we pray and how well we are able to perform, a priest has to be present to perform any of the sacraments? We cannot serve communion, baptize, or even perform the marriage ceremony. They tell us it is to keep us humble and that the Bible (and the Pope) won't allow it. So?"

Angela rarely made such long statements about anything. Moraine was shocked, but as usual she tried to make soothing statements instead of facing what Angela had just said.

Moraine would not forget the change in her own feelings at that moment. Angela was looking squarely in the face of what she perceived as truth.

This is what Moraine had never tried to do.

After hugs and tears, Angela went back to the convent and Moraine could not see her again until her novice days were accomplished. The hugs were a new advance for mother and daughter.

The glacier that had held her in thrall melted another few drops.

CHAPTER 29

The Glamour

Moraine had a bit of 'the gift', 'the knowing', 'the glamour'. Many times she had been able to foretell happenings, some good and some bad. There was no visual image, just a light joyous feeling, or a terrible depressed feeling.

On this April evening, with a heaviness hanging over her, Moraine had an awful feeling of impending doom. At that moment, she looked out her window and saw Brandon running up the street. Brandon was not one to do much running. He had affected a cool, laconic air for so long that now it belonged to him. He was disheveled, hair flying, necktie loosened, breath coming in gasps.

"Ma! Da! Something awful has happened! Angela has been in an accident. They are bringing her home now. Please be home! Please come out! Help!"

Moraine grabbed Patrick from his chair in front of the telly and they rushed out in the street. Bringing her home? Why not to the hospital right away? Why were they wasting time?

Then they could see far down the street where men were carrying a plank that looked like a door with some cloth covering something on it.

It could not be. It would not be. Not their baby. Not their Angela.

They rushed to meet the men. Then, ten paces away, they halted. Clinging to one another, not wanting to go the whole distance where truth awaited them, they stood like stones.

No prayer escaped their lips. Paralysis set in. They knew that Angela was dead but they still hoped for a wound only. The body on the plank seemed so small, just a heap of gray blanket.

They thought she had been safe ... in the convent with the loving sisters. What was she doing here? How could this be? She wore no habit - just jeans and an old black shirt. Where were all those nuns and that Mother Superior who surrounded all of her waking moments?

"Didn't you know?" wheezed Brandon, seeing their blank looks. "Didn't she tell you that she could not abide the rules of the convent? She stayed on as long as she could, using the habit as her disguise but she went out at night to meetings with my Protestants. She was in the middle of a group with me when there was a reprisal shooting. It wasn't her they was after but a stray bullet hit her." Then, as though these words made it real, Brandon broke down sobbing.

Brandon had known, probably everybody in the whole damn town had known, that Angela had turned on her mother's upbringing. Or had she? The convent had not seen fit to let Moraine know. Had they even let Angela use them for a cover? Why? Angela must have convinced them that she was working for a Catholic peace when all the time she was doing the exact opposite. Surely, there had been no clue of her activities. All was confusion.

Moraine had little protective glacier left. She felt exposed to the harsh light of reality.

Slowly, the family returned to the house. Arms linked to keep them together, upright, safe, they took measured steps as though already in a cortege. Moraine took the hated, glacial black shawl and wrapped it around her shoulders, pulling it over her head. She grasped her neglected rosary and began the awful keening sound that started way in the depths of her. Mrs. McBride appeared on the scene telling all and sundry that "sure and I heard the banshee howling last night (was that Tuesday? or Wednesday?) but I could not know who it was for."

Now, Paddy was coming to life. He first directed his anger at Moraine because she had let 'her' daughter into a papist convent. Then, realizing that direction was not exactly on the mark, he began berating Brandon because he had known about Angela's activities and had been part of them. Then, he blamed, in order, the British government, the Irish government, the Irish people (all of them), the Catholic Church, the Protestant churches. Then he returned to Moraine, whom he said had been filling their children with all sorts of violent stories. To his mind, his own apathy and on-the-fence behaviour did not seem to enter into the tragedy. In fact, it would seem to have made little difference anyhow, because the young people mostly listened to other young people.

Moraine stopped her keening for a moment, looked Patrick straight in the eye, and then, right in front of the awakened sleepy-eyed Maeve and Kevin and the stunned Brandon, she picked up her grandmother's fire shovel and hit him over the head.

Patrick was dazed. He was double-dazed. His head hurt and his complacency was hurt. Why had she hit him? Himself, who never harmed a fly? What had happened to his Maureen, the gentle one?

"I've been wanting to do that for years. I'm sick and tired of being ignored, used as a breeding sow, used as a housekeeper, and taken for granted. Don't even open your mouth to me or I'll do it again."

Anger that had no connection with Angela boiled over. She needed something or someone to blame for this irreplaceable loss. She felt more like a volcano that has been pent up for years than like the moraine from a glacier. She wanted to empty herself of all her hurts. In this tragic moment, more of the glacier melted. She became more truly aware of everything around her - colors, touches, smells. She felt as though she were standing aside watching the play unfold.

At this untimely moment, Kathleen rushed in, Angela's only sister. Truly, she had always ignored Angela while attending to her own interests. Nevertheless, her sister, just the same. And on the other side! What would her IRA buddies think? They wouldn't trust her a bit. True to Kathleen's nature, she thought first of how the death would affect her image.

Then it began to sink in. Their family had lost a member, lost forever, and not in the right order.

Kathleen began to know that the horror could have happened to her own children, her own Maeve and Kevin, who were almost the same age as her baby sister, Angela. She started to cry, eye make-up streaking down her not-too-clean face.

Moraine, her rage once again tamped down, had settled down a bit. She felt spent. Her stomach hurt. Her eyes were dry.

'We have spent this holy time blaming each other and ourselves. Is this your gift to us, Angela? Or is this our gift to you?'

Why did her stomach hurt so severely? Was this some kind of reverse labor? Was she borning the spirit of her child back to the other world ... back where she had come from?

When memories came, she didn't try to stifle them as she had expected she would. The rest of the family tried to think of now and the future and to keep their minds off the past when they still had Angela's presence. But Moraine found that the only way she could hang on to Angela was to picture her, remember her, and cherish her. Every ribbon tied, every story told, even the times when Moraine had ignored her and now wished that she hadn't.
When Liam came with his old friend, Father Reilly, Moraine welcomed them with open arms, almost desperately.
Liam had brought with him a gift, a head of Jesus, crudely carved from wood and brought by missionaries from Soweto. He held it out to his mother as they sat with their knees touching. Tears flowed from her eyes. Tears flowed from his eyes and as they prayed together, they looked at the face of the Christ and saw that, in their mingled tears, He seemed to be weeping with them.

CHAPTER 30

Tears and Ice Spears

308 Pender Street
Belfast, Northern Ireland
April 17, 1994

Bridget, Dearest Bridget,
 Have you heard our sad news? Our Angela is dead from gunfire. She really is an angel, now. How can we survive this one? All I want to do is to get back at the fools who think they are saving Ireland - and kill them. That makes me as bad as they, doesn't it?
 I have started reading the newspapers much more carefully these days. I want to know whether or not there is a peace process or just a mess of political gobbledy-gook.
 We Irish are tired of this mess, and now that I see my part in it, I am sickened. I know that my desire for revenge is not the way to respond, but that is still what I am feeling. I thought I had learned not to be vengeful, not to let my feelings explode. It is a long-term holocaust.
 We are stuck in our skins, old potato skins remembering the Famine, remembering the old myths and our history.
 I remember my rage at Paddy when I hit him. I didn't tell you about that? I'm a mite bit ashamed to have lost my so-called self-control. I hit him with a shovel and I am only glad now that I didn't have a gun at hand.
 I'm afraid that all of my frustrations have

181

reacted on my children and grandchildren who now feel that they must have some revenge for everything that doesn't satisfy them. This violence is nothing but a contagious disease and those of us who have 'learned better' had better be working to cure it. But then, I'm not sure I have 'learned better' or what to do with the knowledge.

Do you think that if all women stopped everything that could possibly encourage conflict, that it would make a difference? Half of our world consists of women.

Did I ever tell you about my nightmares? They always come when I have been unable to help some children, sometimes my own.

The night Angela died, I dreamed that a huge window had shattered and it was all icicles instead of glass. The icicles rolled all over me, sticking into my skin and drawing blood. Then they started to melt while I tried to shield Angela, but the blood kept flowing, mixed with water from the melting ice.

This is not much of a letter, I'm afraid, Brid. I can't sleep. I thought I could unload by writing to you and now I find that my baggage is too great.

Please, write and tell me good news of you and your family. Give our love to David.

Your Maureen, (Moraine)

My Dear Moraine,

How can I tell you how I feel, how can I help you through this? How I wish I could put my arms around you and let you cry yourself dry. It's just out of order. A child should never, it seems, die before its parents.

At least you have your own kind around you. It is hard to know what that means until you lose it. Here, we Irish have often been told to "swyok" (stick with your own kind) and stick we have.

Is Paddy able to cope? Men are often so helpless in times of crisis, most only know to turn to the drink. Women, at least, know how to put their aching heads down on the cool tiles in the locked loo and wail. Then, one rises, makes decisions, and takes action.

And the other children... It is my guess that they are all suddenly aware of the shortness of life. I suppose mothers and widows always get the attention and sympathy, leaving the others to deal with their own trauma. Now, when even the older ones need guidance, there is not always someone there to help.

We have offered special prayers for Angela, whom I never even got to see. Candles are lit for her and especially for you, dear Moraine.

Love,
Bridey

P.S. I have been thinking so much about your remarks about women. There is some truth to our power. The questions that come to me are: What will happen to loyalty, just cause, children who must learn to stand up for themselves? How can we live the life we believe we should live?

CHAPTER 31

Wake

The funeral was held just before Easter. The church had been readied for Good Friday and the sense of gloom pervaded the air. Moraine, who had always received great comfort from the sound of the church bells, hardly heard them. She had retreated into the deep recesses of her spirit.

The bells sounded as though they were coming to her underwater and the comforting voices and loving pats went unanswered. Of all things, Mrs. McBride was the most actual comfort. In her enjoyment of a tragedy, she bustled busily about, telling people where to sit, showing them where to go to view Angela, and inviting everyone to the wake.

Later, Moraine would be able to review the scene and everyone in it like a movie, but the anesthetic of the Lord saved her from more than she could take while it was going on. There was a mist before her eyes. Paddy was in even worse condition because he was trying to act like a man, showing none of the wracking sobs that had attacked him in their room.

The wake was a nightmare. Are they always thus? An almost hysterical atmosphere prevailed. Many were roaring drunk - perhaps an escape from their own mortality? All of the wakes that Moraine had experienced before had included the drunks spilling their drinks on themselves and everyone else, but oddly enough, there had been a sense of good cheer. Not so, this time.

At past wakes, one might say, "Good ole Michael has gone to his reward" and in the next breath, someone would remember a splendid story on Michael. There was the hilarious time that Michael had gotten in his old car to drive home without realizing he was in the back seat.

"Thought his steering wheel was stole!" Laughter would peal out. Other memories would surface. But this could not happen today.

The old boys don't sing *Danny Boy* at the wake of a young girl.

The circumstances had been too harrowing. Angela was so young. Smiles only came through tears.

When Gareth, with too much under his belt, tripped on the old faded rug and fell flat on his face, the mourners just parted making a path for his fall, stepping over him on the way to the food table.

Some came to look down on the O'Sheas because their daughter had been meeting with Protestants while in the family of the convent.

The nuns from her order did not come, though some had truly loved Angela. They were afraid of being connected with such a person, of working with the Protestants, of all things!

Sister Bonny came. She moved quietly among the friends and those who had come to look. Often she was in the not inconsiderable wake of Mrs. McBride, calming and soothing. She took Moraine out of the living room to rest a while in her own room.

She did not desert her after all of the shouting was over. She helped her decide what to do with the overwhelming number of cakes that helpful friends had brought. The dinner foods went well, but without a freezer she was up to her neck. They took a whole batch to Angela's convent where they were greeted with stiff smiles and words of condolence that rang falsely from the nuns whom Angela had deserted. Nevertheless, they took the cakes.

Moraine hardly realized the courage it took for Bonny to stay by her. The danger was very present. Among those gathered were members of all factions. All, until they had lifted a few, were glancing around making mental note of who had come to the O'Sheas and trying not to be too noticeable themselves.

Who knew who would turn into *maskfaces* in the night?

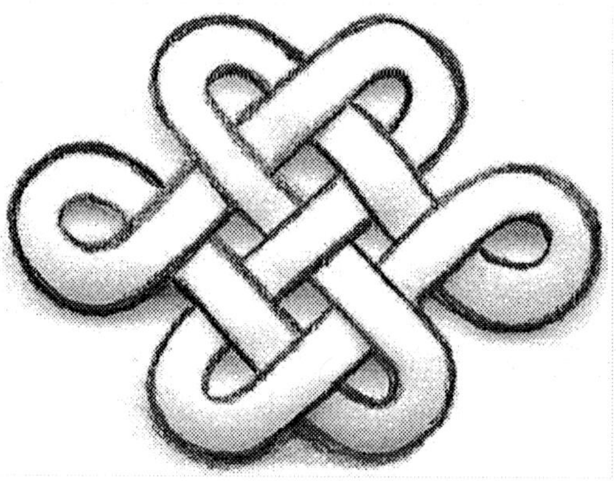

CHAPTER 32

Hooters and Udders

The grandchildren, Maeve and Kevin, seemed untouched on the outside. They much enjoyed the drama of it all, the guests coming to the house, and the wonderful sweets they brought with them. The women vied to bring the best and the showiest of their cuisines.

They wailed politely for the family but it was not necessarily proportional to their sorrow. The volume increased when there were several mourners, each trying to prove that her sadness was the greatest.

In the aftermath, for a while, everything seemed almost normal, whatever that means.

Moraine forced herself to go to market and picked out all the things that Paddy most enjoyed, only the things he loved. Broccoli and rice were *no-no's*. She left on the shelves some of the foods that she herself would enjoy.

Unconsciously, she selected pears and a cheese that Angela had relished. She was in a haze when she paid her check and stumbled to the bus, hardly knowing where she was.

This led her to a positive act, a decision for herself.

'I have many days ahead without Angela. My life will go on, and I am not permitted to waste it.'

Thursday of that week, another icicle melted

from her awareness. She vowed to be totally self-centered for the day. On that day, Moraine picked out the largest bunch of broccoli and headed for the checkout counter where her eye fell on one of the magazines on display. How To Be Sexy, How To Appeal To Men, How To Dress To Attract Men. Those headlines were eye-level with others that told of results of such activity – Baby Expected By Stars, Mother Arrested For Cocaine Use, Golden Winner Declared Worst Dressed – meaning *scantiest* - this from the USA. The rage began to build again in Moraine.

Had all of her efforts of late, to apply eye makeup, tint her gray hairs, do all those exercises for a better bust line, been for these superficial tabloid purposes? What was her real purpose?

Apparently, she had been 'sexy' enough to produce four children after attracting their father in the first place. What were considered *hooters* on the magazine covers could also be thought of as *udders* once the baby process had come about.

The people in the tabloids had one answer, though. When they were through with a spouse, they got out. There were no hidden frustrations for them.

It all spewed out all over the grocery stores counters, the whole story. The women in the stories had such unrealistic expectations, almost the same as the ones she had held when she told her children and grandchildren about how Queen Maeve had taken revenge on her prince. That was the children's favorite and one of hers, too. She loved the way the Queen managed to overcome every time. Maeve was one *prima donna* of a queen, given to dramatic entrances and exits, being fawned upon, getting her own way and being the winner in any way possible. Queen Maeve was a woman truly to be regarded in many ways.

Maybe Moraine could learn from Maeve. Maybe she could learn how to get Paddy back for all the hurts he had inflicted on her. Was Moraine whining, again? Maeve wouldn't stoop to that. There, Moraine had already learned something from Maeve.

The broccoli and rice went on the table and Moraine was the only one to enjoy it. The others nibbled but she relished it because she had done it for herself.

Once again, she registered at Queens University for some courses in literature. The very names of the authors, poets, and playwrights gave her hope that she would find understanding about Ireland. Joyce and Yeats, Synge and O'Casey.

Paddy offered no barrier to her efforts, but underhandedly, he downgraded every educated person they had ever known in a left-handed effort to downgrade Moraine.

In most ways, Patrick was not a cruel man, but he found little purpose to all this learning. Would it earn a quid? Did she think this would make her better than he? Moraine, he opined, must have gotten this phony attitude from her mother who had kept taking courses in everything from quilting to practical nursing.

Moraine, on the other hand, began to feel quite smug about her new controls over her life and her strength in the loss of Angela.

That's when the thought of avenging Angela's death began to edge into her mind.

CHAPTER 33

Vengeance

For a while, Moraine only thought about all the valiant deeds she was going to perform to avenge Angela. Then it occurred to her that she was only fantasizing and that she should either "fish or cut bait" So she started asking around about undercover groups that might be meeting. Her friend Bonnie implored her not to get into this mess because it was not only dangerous but was not a fitting way to remember Angela.

"A fat lot you know about it. You don't know how I feel."

"Oh yes, I do. You are so wrapped up in your myths and legends that you have started to believe them. I understand because I am a woman, too, and we are not always the gentle caregivers we are cracked up to be. We have a memory of what women used to be, fighting for survival alongside the men, a shadow side of us. Please, Moraine. Don't go down this path. It will only push peace away from us."

Dear Bridget:

I've been wondering if you ever got those brass knuckles you wanted. Now, I know what you meant. I have a plan to even the score for Angela.

One of her friends happened to mention in my hearing the name of a group of Protestants that I believe is the one that Angela had joined. I am pretty sure that Brandon is part of the same club and I'm going to get him to take me to one of their meetings. They will probably never trust a Catholic no matter what I say or promise, but it is worth trying. If I can become part of them, I can find a way to get back at the IRA splinter that killed my Angela. Or, better still, if I can belong to that splinter group, I can do damage from the inside. I don't even care at this point which side I'm on. I just want to do damage to those who killed my Angela and at the same time strike a blow at those who have kept this mayhem alive in Ireland. Maybe you had better burn this letter. I thought I was beyond these thoughts but losing Angela has left raw wounds. All of my high-flown ideas about women being able to change the world have dissolved in rage.

    I have plenty of time to plan very carefully. Brandon must not know what I am up to - even if he wanted to help, which he wouldn't, I want to do this all by myself. Besides, he might let it slip to Paddy or, even worse, Liam. Then they would get in my way for sure. They'd probably even get Father Reilly in on the plan. You can imagine the hell he would raise. Everybody wants peace except those of us who have been wounded by the whole process. Peace will have to wait.

    I know I am taking a risk by writing my thoughts of vengeance but I had to tell someone.

Love, Mor.

Dear Moraine,
    The whole mess reminds me of the old feuds in the Appalachian Mountains here in the U.S. Come to think of it, they were probably Irish like

the rest of us. It seems as though a feud never ends because one more person lives that killed someone or one person lives who is related to someone who was killed. Try to remember that Angela was not killed deliberately. She had volunteered to be where she was.

The Civil War, that was such a tragedy to this country, was fought by soldiers who often did not want to fight. They were filled with thoughts of patriotism and were taken advantage of by politics and big business.

Yes, I did get me a weapon. No knuckles. They are too illegal but I abandoned my baseball bat and now a short length of pipe goes in my bag with me and, so help me Jesus, I will pound anybody who tries to hurt one of my family - or me. I have even thought of using it on David, sort of like you did to Paddy.

The thing about David is that he leaves everything up to me - maybe I brought that on myself by managing too well. I'll never be able to fake being a dependent female. I also think I love him.

By the way, Moraine, be very sure that Mrs. McBride doesn't have an inkling of your plans. She could be the very big mouth that you don't need.

                Love and good luck,
                      Bridey

P.S. Where is Kathleen in the midst of all this furor? Is she aware of your zeal?

Dear Bridey,

    A short note to tell you something I can't believe and forgot to mention earlier. Kathleen has tired of all this 'violence', not to mention tiring of Sean. She has a new passion. She is an environmentalist, an ecology freak. She is going to save the earth. I'm not sure I didn't like her better when she was going to blow it up. I haven't seen the signs yet, but there could be a new boyfriend in the wings. You have never heard of such virtue.

                                        Love, M.

CHAPTER 34

Jail

It was about four in the morning, cold and no light yet, when there was a pounding at the door. Moraine grabbed her blanket to wrap around her (unfortunately this left Pat without one) and started to open the door. She need not have bothered. It burst open and five young people in the obligatory ski masks rushed in.

Without hesitation, they ran into the first rooms they came to, the bedrooms, and began to search every corner. They even pulled boards from the floor to be sure there was no contraband hidden there. There had been a tip that the O'Shea's were part of a plot to bomb the police headquarters and they were determined to find the instigators.

They asked about Brandon and his camouflage soldier jacket which was his pride and joy and which was also forbidden for young men to wear. Moraine gave them a piece of her mind, a rather large slice.

"What gives you the right to come in here rousing decent people, scaring little children, telling me about a law that should never have been passed in the first place? You are a bunch of piss-faced toads! Get out of here before I get a gun to you!" Shaking and quaking, she stood between them and her children while the intruders literally tore the house up.

The next day, Moraine dreaded the cleanup that awaited her. To postpone it, she set herself into a frenzy of baking, anything to keep from cleaning. Where was that Brandon?

She kept hearing a rattly sound at her kitchen window. Finally, with her hands all doughy, she looked out and spied Gareth, he of the violent red hair, trying to get her attention by throwing gravel against the pane. When she let him in, he was too breathless with excitement to talk for a moment. Then, gulping and gasping, Gareth whispered, "Ms. O'Shea, the garda got Brandon because he was wearing that camouflage jacket. That's forbidden to us and he got caught. It's skeered I am even to be telling you this. They have taken him to the jail."

To have two such crises in just a few days threatened to unhinge her. In an effort to calm her anger and preserve some kind of normalcy, Moraine had been baking cakes for the church fair. In spite of her good intentions, the sad thing was that some poor, benighted soul would probably buy Moraine's cakes in the name of charity. Since she had sent the last burnt offering, the committee had pondered the possibility of asking her not to bake. It seemed kinder to them just to throw her efforts away as soon as she brought them in.

In addition, Moraine had not been asked recently to sing in the choir. Too much rejection is not a good thing.

The kitchen was always a total wreck when she finished any of her baking attempts and more often than not, the mess would linger for weeks.

Today was different. Today Moraine finally felt strong enough to take positive action against the chaos.

When she heard about Brandon, she very

thoughtfully began cleaning up her mess. She wiped the counter clean of the flour that had spilled, she cleaned the batter from the oven bottom, and she washed her mother's blue mixing bowl thoroughly and put in its proper place. Even the floor that had been getting really sticky yesterday (it would have pulled your socks off) was scrubbed, corners and all. Had one of the family come in unexpectedly, he would have thought he was in the wrong house.

All of this gave Moraine time to fester. Not really think. Fester. The energy she put into the work had helped her make a decision.

She would go the next day. She would go to the prison and get him. She would wrestle every guard that dared stand in her way. She decided to ask Kathleen to go with her in case she needed support. She wanted Kathleen to see what a terrible place the prison was.

In the beautiful morning that dawned Moraine fixed breakfast for everyone, without telling Paddy of her plans. She knew that he might tell her not to go or, worse still, he might offer to go with her. She and Kathleen were totally unable to swallow a bite. Moraine's stomach churned until she thought it was surely visible from the outside. Pat remained oblivious to all but his morning newspaper and for some miraculous reason he skipped the little article inside that told of the arrest of one Brandon O'Shea.

Moraine and Kathleen boarded the bus. Moraine looked at every soul aboard in case she saw someone she knew. She hoped to get through this humiliation without too much ersatz sympathy. She hoped she had avoided Mrs. McBride when leaving the house, no easy task that. Not

199

really ashamed of Brandon, she could not bear the exposure of questions and explanations.

Moraine was in full shake. Everything, externally and internally shook. She tried to light a cigarette but could not steady her hand. She felt like throwing up with nothing to throw. She wished she had not brought Kathleen with her, but she needed support from someplace.

Brandon had been spotted wearing the forbidden camouflage jacket on the day Angela had been shot. That explained why the house was entered and searched. He had been arrested the day after the search and interrogated, then put into prison to await trial. The excuse the police gave was that he had on an army jacket that was illegal for any non-army person to wear.

A prison visit was not what she wanted to do. She could not bear the thought of her own flesh and blood being incarcerated.

She sat on the bus watching a red nosed old sot across the aisle. He must have been coming home from a long night. The smell of stale booze permeated her space, adding to her nausea. Worst of all, the red nosed one whistled over and over. She thought '*I know that song. What is it?*'

The drunk whistled only one line with no inflection. "I'll take you home again, Kathleen." Then he whistled it again and again. No line about where that Kathleen was or where home was. Moraine considered ways in which she could shut him up. She fantasized about it. She considered smacking him across his veined, beefy face. '*I'll just swing my purse across the aisle and shut that old bastard up.*'

Two college-aged girls sat behind her, bemoaning the fact that there would be no jobs for them when they graduated. One friend of theirs had left the day before to teach the hill tribes in Thailand. It sounded quite exotic to them with no notion of the dangers from the tribes.

The bus had rumbled and bumped to the stop where she would wait for the telltale bus, the steel gray bus with the windows painted so no one could see in or out. This would be the bus to take her to The Maze where her own son, Brandon was being held for his 'crimes'.

Would she be allowed in? Did she want to be? Her revulsion included not only the happening itself, but those around who had been participants in the arrest of her son for wearing a camouflage jacket and that was all. They said. Would one of them dare smile? Could she get at the *smiler* with her bare hands?

Her friends had warned her about The Search. She had never been strip searched, although the threat had presented itself at one of the police barricades on the Shankill. Omigod! Will they touch, even touch, Kathleen? Jesus and Mary! Not on my life. Over my dead body!

Now her breath came in gasps. Not only because of the exterior search but also because of the mirror she would have to stand on so she could have nothing concealed anywhere. She wanted to turn tail and go home. Her bravado was gone. But she was committed now. It was, after all, her Brandon.

When the rattling, smelly old bus halted, the passengers were all crowded into yet another smaller jalopy, unpleasantly aware that soldiers with guns ready stood to watch for any infractions. Finally, the jalopy pulled through the last gate and the gates clanged shut. How final that sound! Would she ever get back out? She began to feel imprisoned herself.

Calm acceptance was the face she proffered to Kathleen who was becoming agitated. Smiling, teeth clenched, Moraine sang the same hateful "I'll Take You Home Again, Kathleen" to comfort her. Kathleen's eyes became saucers as she listened to the casual conversation between relatives who had come to visit an inmate.

"Do you know what they did to my brother? I will get them back for that if it's the last thing I do! Dirty Protestants." Or, "Dirty Catholics."

"Now, Darlin'," Moraine told Kathleen, "Remember always to turn the other cheek. Be sweet, now." She knew this was a phony admonition and Kathleen knew it too, from Moraine's tight lips and furrowed brow. Off the bus they climbed.

In all of her nightmares (should that be a female version - maybe nightstallions?) and visions, Moraine had expected her first impression of the huge sprawling prison to be all gray, all different shades of fog. Sad, a bluer shade of gray. Two other very strong images struck her, instead.

First, the noise. Omigod, the noise. The clang of barred doors and gates as they shut so finally behind her, the upturned radio blare, and the obscene shouts from cell to cell reverberated until her very spine joined the rhythm.

The smell. Disinfectant was the overpowering sensation but it could not hide the sweaty body odors or the smell of large pots of gray stuff cooking with a lot of onions.

The color was something else again. In a failed attempt to break the gray syndrome, some creative soul had painted the huge old air ducts what had been perceived as a heavenly blue. The ever-gurgling pipes that ran universally through the whole prison had been painted a bright red, which unfortunately had, due to the heat, become a strange hue resembling dried blood. No one really ever knew what gurgled through the pipes nor did they want to know.

When Moraine got to the inner sanctum, she was amazed when the guards marched (yanked) Brandon out into the room, sans jacket. He had a stubble of a beard, oily stringy hair, and a pungent odor. They literally shoved him at her.

"We have interrogated Brandon O'Shea and find him not guilty of the charges against him." 'Not guilty' does not automatically refer to 'innocent'. It hangs someplace in the middle where 'not guilty' mostly means they couldn't pin it on him this time but that they would keep an eye on him.

Moraine had thought she was beyond the anger she felt now. Her planned speech left her. She was so relieved that she felt as though the wind had gone from her sails.

Weak-kneed, she leaned down and picked up a penny from the floor. Is this really luck? That night the shattered glass nightmare returned.

Brandon never told her anything about his stay in the prison. When he was with his buddies, he sometimes talked to them about it, but always in the recently revived Irish popular with the younger generation. Moraine had forgotten most all the ancient language she had ever learned when she was at home with Granny.

Patrick was never told and if he suspected

all of this activity (and Moraine was sure that he knew), he gave no clue. He had to have known, and acted like an ostrich with his head in the sand. Even his buddies avoided the subject and Brandon was not about to ask for the tongue-lashing he knew he would get. Moraine kept a very low profile until she was pretty sure it was over.

CHAPTER 35

Maskfaces

From that day, all of her spare time was spent trying to find some organization that would fill her needs. She went to any and all open meetings, political or otherwise, often feeling eyes on her back, never really fitting in. She did not aim to fit in, just to slip in. She sat through hours of haranguing talks with conspiratorial whispering going on all around her.

Her impression was that there were a lot of clever people who wanted to show off and keep conflict alive. Still, she did not realize that she was just the same as they, if not worse. Scraps of conversations about a proposed 'kneecapping' reached her ears. Because he lived with a woman from the opposite faith, a man had been targeted for their bullets in his knees, crippling him. All of her married life she had feared that something like that could happen to Patrick because of their mixed-religion family. She had just read in the paper about this happening to a family over in Bangor and the man was left to die. A widow was left alone and nearly helpless with six little kids to raise and both her family and his were afraid to get involved.

Eventually, Moraine attracted the notice of a few radicals (does it even matter which side?) who were scouting for trainees and something about her tense manner made them invite her to go to their next meeting. They were more careful than she imagined.

The maskfaces were only giving test runs to discover the level of trust, reliability, and courage, but Moraine knew none of this. Hopefully, but anxiously, she appeared at the old storefront at the appointed meeting time. The store had been a successful newsstand until it was bombed out about six months ago. Now it was used for meetings such as this. The smell of charred everything still rose to meet the members as they straggled in. In lieu of electric lights, dozens of candles were lit, and a few old lanterns had been pressed into service, giving rise to the speculation that history could repeat itself and the old place could be burned again.

In a tiny room abutting it, there was a table set in the center with a large flag hanging on the front. The colors displayed there made no difference at all to Moraine. She was determined to strike back at **somebody**. A Bible had the place of honor on the table to denote whose side God really is on. A gun beside the flickering candles showed how this group would do its duty. It was a terrifying sight.

"What would you do if the police take you in or the soldiers catch you? Could you keep your mouth shut?"

"What about your family? Would you protect them first or our great Cause?"

On and on the questions came until Moraine was exhausted. She was ready to promise anything even though she was apprehensive about continuing on this path. Then came the big break.

"It's agreed, then. Moraine will take delivery of the boxes. One a day, for ten days. O.K., Moraine, you'll go meet Finn and bring a box home each night and hide it until we are ready for it."

Moraine's heart quailed at the prospect of the real thing. No longer a fantasy. She felt like a right eejit for even thinking she could save the world this way, vengefully. There was Angela to be avenged.

"There's no place to hide boxes in my house. Suppose Patrick finds out. I've changed my mind. Give me a different job to do."

'I am so frightened', she thought but would not dare to admit.

"That's your problem. You said you'd help with the explosives. Just do it. All you have to do is follow this map, keep your mouth shut and for God's sake, don't bounce the crate."

It made her feel no more secure to have the voice coming from a black ski mask. *'That's probably his real face. If he took it off, there would be no one there. There is a whole new race of people with two holes for eyes, one for nose and one for mouth. Just like a child's drawing. Maybe it's a gruff voiced she'*, she fantasized, trying to escape from the dread that threatened to swallow her. There was no turning back. Now that she knew their plans and voices, they could no longer trust her outside of the circle.

She was relieved to leave the smoky, heavy room and go outside for her first caper. Then she remembered the task.

There had never been a night so solid. Palpable. Inky. The time had been chosen so there would be no moon and still there were mysterious shadows cast just behind her, lurking. She was petrified. There had been other women's voices behind those masks, so she knew she was not the only female in the mission of running supplies. Was she the only one so frightened? Instead of the ritual mask, Moraine wore her mother's old black shawl. As much as she had hated the thing, the very smell, the essence of her mother moved her to tears as she started out. It would make her appear to be only a woman out at night rather than one of the rebels.

She sliced her way through the alley, stumbling once over some soft thing, not daring to look back, barking her shins on an old wheelbarrow across the walkway. She had never heard so many rustlings, scufflings and squeaks in the quiet of the night. Were they animals, humans - other? What was that soft thing she fell against? Was her leg bleeding now? She dared not look.

At last, breathless and weak, she came to the place where she expected to be met by a maskface. There was no one there. Once more, she tripped over something and it sent her sprawling - something much sharper, a crate marked with a huge cross barely visible in the murky dark. It was set on an old luggage cart and she realized this was so she could roll it back home to hide. It was her first test. It was the explosives that she was supposed to keep in her house. They had told her that Finn would meet her here. There was no one.

She had a feeling that someone might be standing very near, soundlessly. She could hear only her own gasps for breath.

Staring at the box, it seemed to her to contain all that she had ever learned, a veritable Pandora's box of all the choices she had ever made. There was no time left to reconsider. She had time only to fear to the very bottom of her guts as she rolled that damnable thing back to her own house. Suppose it bumped on some rubble and exploded, blowing her to bits.

This was the box that was supposed to help bring an end to the madness in Northern Ireland. Something rang a false note in her head at that thought but she had to pass over that, remembering Angela's death and Brandon's arrest. She pressed on with the dreaded trek back to the house. Now she had extra dreads. The RUC [Royal Ulster Constabulary] might see her with her illegal load; Patrick might be coming in the house at the same time; she might wake the children.

"This is for Angela," she repeated under her breath over and over. "And for Brandon, too."

Strength comes when we need it. Her adrenaline was pumping furiously as she made her way back to her house.

The attic space of Moraine's house was connected to all of the other attic spaces in the row. She had decided to get the crate up those three flights of stairs by tugging it on an old quilt and then to move the whole thing as far from her own stairs as she should get it. She had blisters on her hands already from pushing the cart up curbs, past the same old wagon that had blocked her way. She pushed very gingerly, past the soft, mysterious bundle lying in the middle of the alley. She was limping now.

Softly, she eased the cart on its metal wheels up the three steps that led into her house, then rolled it on the quilt she had left in the foyer. Most of all, she wanted to sit down and cry before she went on up the steps, but her determination kept her going.

"Where have you been?" Patrick's voice nearly stopped her heart. What was he doing home and awake? It was his card night with the boys.

"I just couldn't sleep," the weak reply was all she could think of and she knew immediately it wasn't enough.

"Don't be trying to pull the wool over my eyes," said Pat, " I know what you are up to, out there causing trouble. What's in that box? Have you been meeting with that bunch of maskfaced killers?"

"Don't you be accusing me of anything. You are just a wimp. At least I am prepared to fight for the memory of my children. You never do anything but march up and down all the wrong streets with your fancy orange scarf and go on about the only stupid battle you folks ever won. What a dinosaur you keep alive! This is the way I can get back at the Protestants that killed my Angela."

There was a heavy loss of logic here. The fact that Angela had been part of a Protestant group was lost on Moraine, who simply felt the need for revenge on somebody on any side for the injustice she felt and she intended to get it. Moraine had about as much interest in the politics of the thing as an ant. Who knows what side a maskface is on? They all look the same.

By now, the grandchildren were standing at the door listening. They had had a hard time dealing with their young Aunt Angela's death and had dreamed of their own revenge. Both were enchanted by their grandmother's derring-do, and both wished they had been in on the 'adventure'.

"Granny, can we help move it?"

"Don't you dare touch a single thing!"

Patrick was stumped. He could help Moraine take the crate up to the attic. Were they IRA supplies? Or were they from that gang that Angela had been palling around with? Moraine needed help with the heavy, unwieldy case. Or should he report her to the police, as an upstanding Irishman should. He couldn't do that to his own wife. What will the children think? There's a quandary for you.

"Pat," whispered Moraine. "I think it might be explosives for manufacturing bombs."

"But for whom? I do hope you have selected Ireland as the focus of your great nobility and not Arabs or Israelis." Patrick was surprised to hear himself speaking so eloquently.

"Well, their women are probably doing the same thing." Moraine was searching for a further answer to this inanity and would surely have come up with something equally ludicrous.

Without warning, all hell broke loose.

Both Moraine and Patrick were frozen in place when the men in the black masks burst in the front door. One, gesturing with his rifle, made Patrick get out on the front steps. Another shot Patrick in the knees. Though he tried to bite his lips and hold back, Patrick cried out in fear and pain.

Who was on which side? All were 'Christians'. All black masks, all Irishmen. What difference did it make? Chaos.

The back door burst open and another batch of maskfaces rushed in laughing.

"You passed the test! You made it back!"

One of them gave the crate a big kick with

heavy boots. Breath caught in Moraine's throat. As the gang looked around, they realized that there were others already there. And that they had shot Patrick!

The second gang started to slink out the same door they had burst in. There had been no plan for confrontation and they had to get out of there.

"Don't do that. The explosives might go off!" Moraine found her voice though it squeaked.

One of the men, on his way out, stepped forward and yanked open the crate. There, all nestled down in sawdust were shapeless chunks of scrap metal - all fake!

In order to earn their trust, Moraine had spent her night in fear for nothing. Now, the thing she had always dreaded had happened, Patrick had been kneecapped for what she had done and for living with a Catholic!

Moraine looked down and saw that there was blood oozing from the small of his back, in addition to his knees. In response to Moraine's hysterical cries for help, the neighbors called an ambulance. Patrick was taken to Queen's Hospital where he spent some painful time in recovery.

Mrs. McBride, of course, was much in evidence actually helping with the rest of the family.

Sitting in her old green chair, wrapped in her mother's shawl, with Beltane and Samhain stroking her legs with their furry bodies, drinking tea from her same old cracked flowered cup, she could finally see.

She saw that she had made things worse when she should have been in some action to resolve The Troubles.

Moraine's scalding tears melted the rest of the glacier that had covered her awareness of the universe. She could see very clearly now.

Oh, what a tangled knot we weave ....

Epilogue

Three Years Later

Three years after Patrick came home from the hospital, Moraine was still struggling to make a life for herself.

Patrick was paralyzed from the waist down but his arms and trunk were still powerful. His natural good humour had not totally forsaken him altogether but he now had periods of depression. Moraine thought this would be her chance for atonement. If she could take full care of Patrick, she would make up for her blindness. It was too much for her to maintain.

'Oh, for the Lord's sake, Pat, snap out of it'.

She would catch herself, bite her lip, and try to hold back her rush of tears.

Just when she thought she would not be able to stand it a minute longer; the meals, the cleaning, and the children, a sort of miracle occurred. Brandon came home.

Moraine had been so afraid that she would have to deal with Mrs. McBride's curiosity that she had arranged for Brandon to spend some time volunteering with a respite care facility in nearby Warrenpoint. He had his bed and board there.

Without explanation, he pitched in with his father's care - the toileting, the shoulder to lean on, and the lifting. The real saving grace appeared in all the man talk that went on about sports, the sharing of a brew and the spinning of a tale or two.

Moraine was left out. Just when she had decided that martyrdom became her, just when she had all her plans made to show how she had changed, she wasn't even needed. It took the wind out of her sails.

But it was to Moraine that Patrick turned at night, as they lay in bed, sleepless. It was Moraine who heard how frustrated his paralysis had left him.

"Right in the prime of me life. I have so much left to do."

From the earlier time when Patrick had not being able to tell her about losing his job, he now went to the other extreme, telling her all of his feelings. In her pain, Moraine thought it was only her due to be the sounding board no matter what it took from her.

Her impatience with Brandon was somewhat less controlled. His only job was part time at a fast-food restaurant. Even in the face of his being practically indispensable in the care and well being of Patrick, it took more understanding than she could muster. She could not find anyone but herself to blame for the circumstances in which they found themselves.

One rainy morning, it dawned on her where her misperception lay. She waited restlessly for Sister Bonnie to put in her regular appearance for morning coffee. She even dusted. Her new plan was to advertise for young mothers to come to a schoolroom on Saturdays to share their experiences and to learn to take steps to begin to alter the situation. She would wave her magic wand and in one fell swoop, she would change the whole scene to one of peace.

Kathleen, who now only left the children with her mother while she went to work on the environment to save the earth, would oversee a day nursery, (even if Moraine had to wring her neck).

What steps would she propose? Where would she get money for expenses? How would she get the word out to potential mothers?

Moraine and Bonnie sat head to head deciding how to begin. Of all unlikely allies, Mrs. McBride arrived on the scene with a little nest egg that she wanted to contribute.

"I have wanted to do something for the children (and their mothers (or even grandmothers (or just for Ireland))) for a long time. I just didn't know where to start, " she said in her usual manner of interjected speech.

The programs would have to be interesting. Aerobics, diets, and childcare could be programs if they could find volunteer teachers and free brochures. The underlying reason for the gathering would be Moraine. For ten minutes during each session, she would share her realizations about what power women really have for influencing the future. She would start, hoping a discussion would evolve. She would tell them that they would be the catalyst for a change to peace and show them why they must exert care in raising their families.

Each detail was arranged and when the Great Day arrived - guess what - nobody came. Moraine was devastated. She had been holding such glorious visions of herself – well dressed and in full control - and a vision of her guilt being wiped out in one fell swoop.

"Ma, they are just afraid. Don't you remember the meetings in Belfast and Armagh?" said Kathleen soothingly. Moraine was all for quitting the entire project but Bonnie would not agree.

"Step by step, Mor, step by step. We thought we could wave a magic wand and it would be done. We'll keep nibbling away at it until we succeed."

All well and good but it was frighteningly apparent that Moraine would need a paying job since Patrick was disabled. She had planned on her new project becoming such a success that she would be able to go to work without worrying about changing the world. Now, she planned to work in a school or some trauma center and reach large numbers every day with her message. Her dream included quite a bit of recognition for her nobility, which she would receive modestly.

She had no training. Moraine was an intelligent woman, particularly good at organizing other people; she had begun with that stint in the creche before she was married.

It is well nigh impossible to convince prospective employers of such talent without a resume. To her combined disgust and relief, she finally found a position in a large department store much like Marks and Spencer's. She was assigned to the infants' department. That was about the last place she wanted to be. She could see no way that she could aid her Cause in such an environment.

She was wrong. Fate had a different role for her. Along with her hopes for the failed Saturday project, she now had access to all of those mothers she wanted to influence.

She began her 'ministry' right in the midst of bassinets, baby baths, and fluffy little dresses. She buttonholed women who would rather be selecting a fluffy toy and warned them about their inner frustrations that could influence their families. She was a royal pain in the neck but about one in every ten women paid attention. She overdid it, causing antagonism that almost lost her the job until she learned to turn down the volume.

She took names and addresses and assured each one that she would find a site and time for the get-togethers and notify them. Moraine's passion was so deep that she would not let go. She was a puppy chasing its tail. Her own perspective shifted a bit from trying to forgive herself to finding a genuine desire to help these young women avoid the trap of aiding violence without realizing it.

"But I need to know about health insurance. Suppose something happens to my husband."

"I don't want him to know I'm going to a meeting. It just upsets him when I seem to know as much as he does."

'Aha!' thought Moraine, *'Straight into my web.'* Then and there she signed Mary and her friend Teresa for her first gatherings.

When her thoughts went from herself as the center of the grieving universe to those who might have a bearing on the future, Moraine felt her dreams begin to fall into place. It happened slowly, to be sure.

Sunday afternoon became the time for the meetings and Father Liam was able to provide the rattly old van to bring the more frightened ones into the schoolroom. In spite of all her efforts, there were never enough attendees to satisfy her. Sister Bonnie and Mrs. McBride tried very hard to convince her that she should be grateful for those who came but Moraine had the soul of a crusader.

Patrick was impressed into addressing cards of reminders of dates. He was not pleased or at least he said he wasn't. He called the whole project 'the woman thing' and tried to demean it. He was a phony though, because he was proud of Moraine for trying at all and he was so grateful that there was something he could do that felt constructive.

In the process of doing the paper work for the 'Woman Thing', he met a young woman whose husband had a rapidly growing plumbing business. There was an immediate need for a switchboard operator and the operator could work from home. Patrick jumped at the chance, had the equipment installed in their bedroom, and went to work. What a triumph to receive a paycheck again, albeit a small one!

Brandon was relieved from his daily chores to the degree that he had only to get his father underway in the morning and then he was free to go to work. He considered this to be a mixed blessing because he had enjoyed the growing friendship with his father and he was not anxious to be pushed out of the nest. He had been an accountant before and, *By God*, he could do it again. Moraine gave him more unconditional support than she ever had before.

Through Moraine's new contacts, a job was found for him. She had learned that she needed to be more loving to him, to show him that Irish pride was not all he needed to advance in his life. That was another slow process, certainly not an overnighter.

Her grandchildren remained as one of her deepest concerns. Because they had witnessed the awful night, she felt duty-bound to spend time with them to help them understand and tell them stories that showed peaceful endings even though the dragon had to be exorcised. Moraine knew that the Trauma Centre in Belfast used storytelling for whole families to ease the pain and help understanding.

Here again, Patrick helped. He would tell them, at the drop of the hat, about his night. He held back nothing; let them see what the consequences could be. He showed how he believed that all of Northern Ireland was suffering those same consequences.

Moraine's entire glacier was long gone now. Her awareness was both painful and joyful as the new life began. She touched so many more people than she even imagined as she went through her exciting new life.

Moraine never received a medal for her loving work. She was never the subject of a feature in the newspaper. She did not save Northern Ireland, but she made a difference.

>308 Pender Street
>Belfast, Northern Ireland
>October 16, 1998

Dearest Bridey:

Have you been listening to what I'm telling you? Do you realize that the same culture has influenced us, not to mention all women? I have met some women from other countries and they are making the same mistakes we have made with our families.

I am exhausted from my driving passion but it is worth it. You were so right when you advised me to get out of the house and see what the world is doing. Washing dishes and dusting is not enough.

I thought I could ride off into the sunset in a cloud of glory, completely successful, honored by all. Bonnie and Mrs. McBride keep reminding me that it is a step-by-step process of loving and awareness. Of course, Mrs. McBride always tells me in parentheses which seem to go on and on. I have learned to value her, though, and maybe that is enough of a miracle.

Love,
Moraine